Recording Spaces

I dedicate this book to the memory of my two grandfathers, Dick[1] and Wally[2], from whom I no doubt inherited so many of my characteristics

---

1 Dick (Richard) Chapman: professional footballer for Blackburn Rovers, Blackpool, Stalybridge Celtic and Clapton (Leyton) Orient; who was also a lover of the visual and performing arts.
2 Wally (Walter) Newell: by all accounts a 'first edition', but who had a great affinity to plants and animals.

Recording Spaces

Philip Richard Newell

Focal Press
An imprint of Butterworth-Heinemann
Linacre House, Jordan Hill, Oxford OX2 8DP
225 Wildwood Avenue, Woburn, MA 01801-2041
A division of Reed Educational and Professional Publishing Ltd

ℛ A member of the Reed Elsevier plc group

OXFORD BOSTON JOHANNESBURG
MELBOURNE NEW DELHI SINGAPORE

First published 1998

British Library Cataloguing in Publication Data
A catalogue record for this book is available from the British Library

Library of Congress Cataloguing in Publication Data
A catalogue record for this book is available from the Library of Congress

ISBN 0 240 51507 2

Typeset by Avocet Typeset, Brill, Aylesbury, Bucks
Printed and bound in Great Britain by Biddles Ltd, Guildford and Kings Lynn

Contents

About the author

Philip Newell began his career in the music industry after leaving school in Blackburn, England, in 1966. His first job with live music was as an assistant sound and light operator in a number of 'Mecca' ballrooms in the north of England. In 1968, he moved to London, where he then worked as a live sound engineer in some of the country's largest ballrooms. In those days, touring bands did not travel with their own sound reinforcement equipment, but used the in-house equipment of each venue. During this period of time, he was fortunate to work, in some cases many times, with artistes such as The Who, Booker T, & the MGs, Sam and Dave, Jnr Walker and the All Stars, Wilson Pickett and many other 'classic' musicians from that era. This was an exceptionally fortuitous beginning, because he realized at a very early stage that no equipment, no venue, nor any special technique could make up for pure musical talent: a topic which threads its way in and out of this book. Stars shine!

In 1969, Philip became more involved in recording, and was fascinated by the concept of studio design. In 1970, he designed Majestic Studio, in south London, then moved to Pye Studios, Marble Arch (London), at the end of the year, where his background in live music made him a prime candidate for work with their mobile recording unit. This was still a golden era for music, and in twelve months with Pye he was involved in the recording of rock artistes such as Traffic ('Welcome to the Canteen'), Stephen Stills, The Who, The Faces, The Grease Band, Emerson, Lake and Palmer, and many others. During the same period, recording work also involved the English brass bands, Welsh male voice choirs, Scottish pipes and accordians, church organs, fairground organs, musicals, and classical piano recitals. At the end of 1971, he was invited to join a fledgling Virgin Records organisation as a chief engineer of their first studio, The Manor, in Oxfordshire.

Again, there was a great intensity of work with well-known artistes, including The Bonzo Dog Band, John Cale, Fairport Convention and Elkie Brooks, but Philip felt restricted in the studio, and wanted to return to his travels. On New Year's Day, 1973, Richard Branson offered him a partnership in a new mobile recording company, if he would commit himself to it 100%. As that was what Philip wanted to do anyway, the discussion on the subject was very brief. The first Manor Mobile, designed by Newell himself, began work in July 1973, and did what was perhaps the world's first recording with a 24-track mobile vehicle in the following month. Some of those recordings are on the CD 'Going Live etc' which is still available almost 25 years later.

Once again, Philip found himself in a fortunate situation, with many of the world's leading artistes gravitating towards what was, at the time, Europe's most advanced mobile recording vehicle. The list of recordings in which he was involved is far too long to publish here, but his widespread work included such a variety of musical styles as Captain Beefheart and the Magic Band, Tony Bennett, Queen, Dizzy Gilespie, Alvin Lee, The Duke Ellington Orchestra, Jack Bruce, The Warsaw Philharmonic Orchestra, Gary Glitter, Johnny Halliday, Tangerine Dream, The London Symphony Orchestra and, of course, Mike Oldfield, with whom Philip engineered or produced six albums. These included 'Exposed', the live recording, made from concerts in nine different countries, of an entourage which included 45 musicians, 40 road crew and three PA systems – one for the rhythm section, one for the strings, and one for what remained. This extravaganza took place in 1979, and as well as producing the album, Philip Newell also did the live sound for the concerts.

By 1978 he was technical director of the entire Virgin recording division, and was charged with the choice of design and the overseeing of the construction of the now famous Townhouse Studios in London. It was direct experience of the great and different sounds that could be achieved in non-studio locations that drove him to break with tradition and build the famous stone room in Townhouse Two.

Although Philip had been very much involved with the early Punk Rock scene, recording artistes such as the Buzzcocks, Siouxie and The Banshees, Wire and even later bands like XTC, by 1981 he had become a little disillusioned by the whole Punk/New Romantic era. This disillusionment, together with a faltering romance and an urge to spend more time flying seaplanes, led him to sell his shares in Virgin in 1982. The seaplane saga began in 1977, when Richard Branson bought Necker Island, an uninhabited island of 84 acres in the British Virgin Islands. Newell had been given the task of preparing the island as a tax haven recording studio, and seaplanes were the best means of travel to its deserted shores. However, on Margaret Thatcher's rise to power in 1979, her tax reforms slashed the 93% rate that many top artistes were paying on UK recordings, and the raison d'être for the island studio ceased to exist. Branson built a luxury island resort instead. For a period of six years, from mid-1977 to mid-1983, Philip Newell operated a seaplane company in the UK, and piloted aircraft for air displays, films and television, and he also became an instructor of pilots.

The wind-down of his involvement in music from 1982 to mid-1984 gave him time to reflect on many concepts of music and studios, and when in mid-1984 he returned to studio design, it was with a refreshed approach. From 1987 to 1991, he sponsored work at the Institute of Sound and Vibration Research, in Southampton University, both on aspects of loudspeaker performance and room design. From this work, he wrote many articles and technical papers, and this prompted the writing of his first book, *Studio Monitoring Design*. Although acoustic design is now his main occupation, Philip has never completely left the realms of engineering and production. He still makes live recordings, and in 1995, at the football stadium of

'Sporting' in Lisbon, Portugal, he recorded (with help) nineteen bands and five hours, for television and CD use, using a discrete 104-track recording system, by means of thirteen synchronised 8-track machines. This was certainly one of the largest recording set-ups ever used in Europe.

Philip Newell is a member of the Institute of Acoustics, the Audio Engineering Society, British Mensa, the US and Canadian Seaplane Pilots Association, and the League Against Cruel Sports. Since 1991, he has lived in Spain, which he now considers to be his home, though he continues to travel the world in the course of his work. To date, he has worked in 33 countries, Ukraine being the last one added to the list, where he is currently designing a concert hall.

Acknowledgements

I would like to give my deepest thanks to Janet Payne, without whose encouragement and enthusiasm for the project this book would almost certainly never have been written.

Thanks also to Dr Keith Holland, research fellow at the Institute of Sound and Vibration Research at the University of Southampton and lecturer in electro-acoustics to Tonmeister students at the University of Surrey, England. Aside from his calculation and generation of a number of computer figures in this book, I would like to thank him not only for providing the Glossary but also for the many hours of discussion, and help in ensuring that whilst this book aims to be accessible to as wide a readership as possible, it remains technically accurate. In acoustics, an 'accurate generalisation' tends to be a contradiction in terms.

Finally, many thanks to Sergio Castro, who spent countless hours at his computer, translating my hand-drawn sketches into the constructional figures that appear in this book. I was fortunate to have a friend who is not only artistically inclined, but who also understands the acoustic principles being described.

Philip Newell
Moaña
November 1997

Introduction

In many countries of Europe the title 'engineer' carries a degree of respect. Indeed, it is equivalent to a university degree. Recording companies in these countries are reluctant to call what in English would be called recording engineers by the equivalent in their own languages, but tend to use words that translate as 'technician'. This can be a little confusing when translated back into English, as 'technician' suggests more the role of a maintenance engineer. The companies are reluctant to verbally elevate the 'technicians' to 'engineers' partly because they fear demands for an engineer's wage. Frequently, the maintenance people are paid more than the recording staff because they are seen to have more recognisable 'qualifications'.

The reason why I begin here is that it serves well to highlight the way in which a modern industry has developed from its earliest days. When recording began, it was very much in the hands of scientists and engineers, and the studios were usually staffed, literally, by 'men in white coats', all highly specialised in their respective areas of work. These were the people who dictated almost absolutely how a studio should look, sound and operate. These people *were* engineers, in the full and true sense of the word, but they were frequently detested by the musicians, who often complained bitterly that *their* needs were not understood.

It was during these times that the term 'recording engineer', or its equivalent in other languages, was coined, and as the industry grew, the title remained. It has, in more recent times, largely been inherited by what frequently used to be referred to as 'balance engineers' – the people with their hands on the mixing consoles. In fact they were in many instances the *only* people with their hands on the mixing consoles, as trades union demarcation policies usually prohibited any other person from operating the console. It was all a very technical process, with musicians rarely entering control rooms. The artistic side of the work was normally confined to the other side of the control room window, with only a producer representing the musical side of the process in the control room itself. Musicians played, engineers recorded, and, all too frequently, a considerable gap existed in the understanding of each other's needs.

Gradually, this gap was closed, usually by some of the engineers with a greater musical feel being more warmly received by the musicians. Comfortable musicians usually performed better than uncomfortable ones.

Starting in the 1950s, but developing more in the 1960s, there emerged a breed of 'name' recording (balance) engineers, who built up strong rapports with their artistes, and worked for them on a very personal basis. At times this proved difficult, as most recording studios still prohibited recording by engineers who were not part of the permanent staff, but there were always 'compliant' studios, where these new relationships could develop. Finding compliant studios with good rooms was another problem, as almost all studios were designed by architects and engineers, and complaints were often heard from the musicians that they were not adequately consulted. Almost everything in a studio was a result of technical requirements, and it was not until the late 1960s that studios built for, or by, musicians were anything other than rare exceptions.

Unfortunately, however, many of the studios built for the musicians' comfort were not well received by the engineers, who were mainly brainwashed into believing that certain technical requirements were paramount. The happy compromise was still the exception rather than the rule. By somewhere around 1980, when recording technology and musical instrument technology began to blur, the role of recording engineer frequently became usurped by people who were primarily musicians, or who certainly had considerable musical ability. At last the cultural divide between recording engineers and musicians began to dissolve, but this new generation was normally more electronically competent than acoustically competent. Music recording was entering a world of electronics, computers and digital signal processing. A great proportion of musicians and recording staff alike were drawn into this new domain, and for a time, acoustic recording spaces almost fell by the wayside, except for orchestral recording, film scoring and a few die-hards in other areas.

Studio design has rarely been a comfortable companion to the recording processes, with either outdated attitudes or more advanced technology always seeming to thwart a comfortable evolutionary process. Of course, throughout all this, there have been the small number of people with the perception, skill and foresight to turn adversity into advantage, or to be able to select the most appropriate facilities for each recording. The music industry, however, is now far too important in terms of employment and many national economics for it to continue to rely on the special few for its continued existence. Especially now, as we are beginning to emerge from a period where many people thought that electronic processing would provide the answers for all acoustic problems, a wider understanding of recording accoustics is badly needed. We need more good acoustic spaces, and these need to be built before too much of the art is lost.

The aim of this book is to discuss some of the concepts of recording room acoustics; why they are needed, how they are achieved, and the bearing that they have both on musical performance and recording techniques. However, the first chapter will deal with the sound isolation shells, or 'soundproofing' in everyday language. As we will see, the soundproofing and internal acoustics regimes overlap to some degree, but it seems that no matter how much time passes, the two get very much confused. In June 1990, in the UK publication *Home and Studio Recording*, I wrote an article entitled 'The Great Egg-Box Fallacy', in which I tried to point out that sticking a few acoustic panels on the wall (or egg-boxes in the old days) will do absolutely nothing to stop the sound of a drum kit from penetrating from one room to the next. Slowly and surely, and with many people writing articles making the same point, the message is beginning to get across to the multitude of 'home studio' operators, but outside professional recording circles (and of course those of academic acousticians), the full differentiation between sound proofing and internal acoustics regimes is still frequently missed.

Chapter 1

Isolation shells

There is no space that absolutely cannot be used as a recording studio, so long as it is not subject to any permanent extraneous noises which would intrude upon the recordings. This fact is well borne out in the 1990s by the number of garages, bedrooms, attics, garden sheds and other spaces which *are* used as studios, sometimes even commercially. So why should we go to great lengths to design and construct a purpose-built studio? Well, the above spaces may suffice for a restricted type of recording, during limited periods of time, or with a limited range of musicians, but they are not usually suitable for high quality work over a wide spectrum of music and instrumentation.

1.1 Basic requirements

For truly commercial operation, a studio needs to be able to accept recording work of varied nature, and usually needs to be able to record 24 hours a day. The studio should also be capable of working in an efficient manner, without disturbing any neighbours, and without the neighbours disturbing the recordings. Efficiency in terms of time is crucial, because the client may be paying engineers, producers, roadies and musicians by the hour. Delays can be expensive, as the whole team needs to be paid even if the inefficient studio grants extra recording time free, in lieu of time lost. However, more importantly than all of this, musicians *cannot* be kept waiting if their performances are not to suffer. Delays caused by adjustments for technical reasons, or for disturbances by external noises, can be totally ruinous to the creativity of musicians.

1.1.1 Isolation

One of the prime needs for a viable recording studio is to have a space which is acoustically isolated from the outside world to the degree of about 90 dB. Drums and electric bass, playing loudly in a studio, can achieve levels in the order of 120 dB. A quiet bedroom in the country may have a background noise level of 20 or 25 dBA from a gentle breeze outside, distant traffic, cows moving around in their sheds, or whatever else. In most countries that I know of, legislation does not permit studios, discotheques, or any other 'industrial' sources of noise to more than subjectively double the background noise level for any of their neighbours. At mid-frequencies, a subjective doubling of loudness represents an increase of 10 dB, though at frequency extremes, this can be less. In order to play safe, some countries or local authorities will not permit an increase of more than 5 or 6 dB in background noise and some authorities even none at all. However, 5–10 dB on a background noise of 20–25 dB usually renders a situation whereby, if we can avoid raising the background noise level of our neighbours to more than 30 dBA, all will be well. Subtract this allowable 30 dBA at the boundary of our neighbours' premises from the 120 dBA of the bass and drums in the studio, and we encounter the aforementioned 90 dB isolation requirement. (Readers unfamiliar with the dB and dBA scales should refer to the glossary at the end of the book.)

I use the term 'subjectively doubling' in the above paragraph because, in psychoacoustics, subjective and objective assessments very frequently do not closely correspond. In Chapter 6, Figs 6.1 and 6.2 show curves of subjectively equal loudness, spaced apart by the distances of the perceived doubling of sound; that is, the step from any one curve to the next, at any given frequency, represents a perceived doubling (moving up the charts) or halving (moving down the charts) of the loudness. It will be seen that not only do the curves generally not correspond to any objectively consistent doubling, but neither do they show much correlation between different frequencies. Nuisance noise, also, can be perceived to be very different, subjectively, between levels which may have been measured to be objectively similar. Rhythmical sounds, for example, can often be annoying at levels of only a quarter of the power of random noises. It should therefore be emphasised at this early stage that subjective perception may differ considerably from objective measurement.

1.1.2 Siting

It is true that we could site the studios in remote locations, but even if we take earthquakes as an exception, wind, rain, hail, thunderstorms, dogs barking and other country noises can still require considerable isolation if acoustic recordings are not to be interrupted. Of course, there is always the lucky break, such as a space in a building which is surrounded by book depositories, which neither produce noise nor are sensitive to it. These are rare exceptions, but in any case, in industrial areas, such circumstances are somewhat unstable. It may be that a year after building the studio, the book company turns its store into a printing room, with noisy presses which could disturb the studio, or, on the other hand, they could even turn it into a proof-reading room, where the readers would be disturbed by the studio noises.

Choice of location is an important factor, as to site a studio close to a 'permanent' book depository would be far more prudent than siting it next to the reading room of a public library, which would simply be asking for problems. In mainland Europe, however, there do seem to be an alarming number of studios on the ground floors and/or in the basements of apartment buildings, which are frequently the only available commercial spaces at affordable rents. Sometimes one cannot have everything, and in these cases, compromises must be accepted. Nevertheless, there are some seemingly obvious locations which should be avoided. Sites near underground and overground railways, together with sites under low-level aircraft flight paths or near to busy roads with fast, heavy traffic will all be likely candidates for low frequency (LF) isolation problems. As all of the above examples produce relatively high levels of low frequencies, and these are the ones most difficult to keep out of the studio in terms of cost-effective isolation.

Locations close to police, fire and ambulance stations are also best avoided, not only because of the potential noise problem from vehicle sirens, but because they often house powerful radio communication equipment, capable of creating many radio frequency (RF) interference problems in sensitive recording equipment. A little careful examination of the surrounding area before a building is chosen as the location for a recording studio can save a great deal of cost and trouble at a later date. Even if the technical staff of a proposed studio are confident that they can adequately deal with the RF problems in their equipment, this still does nothing to prevent problems occurring with the electric or electronic musical instruments which visiting musicians may bring to the studio. I recall one well known guitarist being unable to use a very expensive studio because his favourite guitar was exceptionally sensitive to the electricity company's data transmissions along some nearby, overhead, high-voltage power lines. I have also known serious problems to develop in studios close to railway lines which have, long after the studios first opened, installed digital signalling and control equipment. In some countries, state-owned railway and emergency services have immunity from prosecution in such cases, so little can currently be done to remedy these problems.

When I worked for Pye Studios in Marble Arch, London, in 1970, there always existed some problems from the rumble of underground trains, deep below. Occasionally this caused disturbances, but in those days, with the troublesome frequencies being so low, and analogue equipment having little response below 40 Hz, it rarely stopped the recordings. However, further along the same street, the studio Recorded Sound (later Nova Sound) was situated on the ground floor of a hotel building, and operation beyond midnight was prohibited. Originally this was thought to be no problem to business, but those involved in delayed or urgent sessions found it severely limiting if the midnight curfew stopped an unforeseen necessary extension, and bad memories of such problems could lead producers to book other studios in future. Neither of the last two studios mentioned exist today, and indeed it is probable that Pye could not have survived into the digital age without much better LF isolation, given the lower frequencies which can be recorded on modern digital equipment. On the brighter side, one London studio, which I regularly visisted, was built in the Lady Chapel of a church that had largely been destroyed by bombs in World War II. This was a perfect choice of location, in a building with massive walls, situated in its own grounds and away from any main roads or railway lines. Construction was relatively inexpensive, results were good, and there were no restrictions on its hours of use. The owners' choice of location was very prudent, especially since the rent was also reasonable. This good choice of building ensured that construction and operating costs were kept down, so the studio could offer excellent facilities, with good access and few operational restrictions, at an affordable price. Consequently, it has been able to survive some national financial recessions which have forced many more expensively constructed studios, or ones with more restricted usage, to close.

1.1.3 Structural considerations

Not only is siting important, but so is structure. The aforementioned studio in the old church was in an excellent building, but so many of the reinforced-concrete framed structures, very common in southern Europe, resonate intolerably. Reinforced concrete, when used without damping additives, is an excellent conductor of sound. Once sounds get into such structures, they travel throughout the walls and floors with remarkable ease. Buildings made from blocks, interspaced with layers of dirt or cement, meet changes in characteristic acoustic impedances as the sound travels through the different materials of the structure. This leads to losses of acoustical power in the same way that mismatched impedances in an electrical chain would reduce power transfer. On the other hand, poured structures of reinforced concrete produce homogeneous structures, without interspersed materials, so sound is conducted through them very easily. The only simple way of reducing this problem is to add damping materials, such as Concredamp, to the concrete before it is poured, which, of course, must be done during the construction of the building, so it cannot help with an existing structure.

1.1.4 Power transfers

So, careful choice of site and structure can make life much easier for a studio designer and less expensive for a studio owner, but, in almost all cases, some sort of isolated structure will be required if a very high degree of sound isolation is to be realised. Ninety dB of sound isolation represents only one part in a thousand million (10^{-9}) of the sound power in the room escaping to the outside world. Even 60 dB of isolation will allow only one part in a million (10^{-6}) of the sound power to escape. These figures help to put into perspective the degree of isolation work which may be necessary.

As explained earlier, one of the main benefits of a heavy, stone-walled building, with dirt infill, is that there are differences in acoustic impedances between the layers of construction materials. Most people reading this will probably be aware of impedance ratings for microphones and loudspeakers. Connecting a 10 K ohm microphone into a 600 ohm input will not provide a good electrical power transfer of the microphone output to the amplifier input. The connexion of a 16 ohm loudspeaker to an amplifier output rated for 4 ohms will not be able to employ the full output power capability of the amplifier. For maximum power transfer, the source and destination impedances must be matched. The effectiveness of most sound isolation techniques thus involves *mis*matching of acoustic impedances, to produce a series of power *losses* as the sound passes from one material to another.

Some knowledge of these impedances is required for a design to be effective, as impedances do not always follow obvious physical properties. Many people will recall seeing films of submarine warfare, where the submarines have to remain as silent as possible, with all motors and machinery stopped, when being chased by destroyers. Typically, on a destroyer, there would be sailors wearing headphones and listening via hydrophones, lowered into the water, for any sounds from the submarine. A dropped screwdriver, or other hard object, in the submarine would rapidly give away its direction to the pursuing destroyer. The reason why the submarines had to remain so incredibly quiet in order to avoid detection is because the seawater and the steel of the submarine hull have almost identical acoustic impedances, so the sound transfer from one material to the other was highly efficient. Modern submarines are now usually completely covered in rubber, the acoustic impedance of which is very different from steel or water. The sound passes inefficiently across the steel/rubber boundary, then equally inefficiently across the rubber/water boundary. The ease with which a destroyer can listen in to the inner sounds of such submarines is thus greatly reduced.

1.2 A typical isolation structure

This book is not intended to be a textbook covering all situations of sound isolation, so suffice it to say that the variables in this area are very great, and the techniques for dealing with them are multitudinous, so we can only deal with some examples here. Nonetheless, with what follows, we should be able to gain a good understanding of many of the problems in general, and also gain a feel for some ways of solving them. Precisely how to begin designing an isolation shell depends on many factors including the available space, the nature of the structure, and the strength of the floor. On the fourth floor of a building with a wooden frame, for example, it would be unlikely that the weight of a concrete-block wall structure would be supported, so other means of isolation would need to be adopted. However, we cannot continue our discussions about room acoustics unless we have a suitable room to begin with, so let us go through the structure of one option in terms of its isolation shell. Let us suppose that we are on a ground floor, which is a very common situation, and one in which a concrete-block wall *would* be feasible.

1.2.1 Isolation walls

It was stated earlier that dirt-filled, stone-block walls are usually good for isolation. Well, we can conveniently simulate such a structure from more convenient materials by building our walls from high-density concrete blocks of the hollow variety. These are frequently made in sizes about 50 cm × 20 cm × 20 cm, with central dividers. They are closed on all surfaces but one, and can easily be filled with sand. If we build such a wall inside our building, with an air space between this wall and the structural wall, we can achieve an effective start to our isolation process. The concrete/sand/concrete cross-section of the largest surface area of the wall creates an impedance mismatched sandwich, which is both heavy, and highly damped by the friction of the sand particles. This damping is easily demonstrated by tapping an empty block with a hammer, whereupon a clear 'ring' will be heard. Once full of sand its response to a hammer blow will be decidedly dead. The sand should, if possible, be dry, as wet sand will take months to dry out and will risk spreading its dampness into the rest of the structure. The great weight (mass) of such a structure is an important factor in the effectiveness of its low frequency isolation.

The space between our internal and external walls is also of importance, as the air space provides another impedance 'barrier' which must be crossed. In general, the larger the air space the better the isolation, but studio owners rarely like to not be able to see the space which they are paying for, so air gaps in general tend to be small. It should be noted here that with greater air gaps, lighter wall structures can often achieve similar isolation, which is one answer to the floor loading problem in lightweight buildings, but the price to be paid is in the loss of space. The effectiveness of a smaller than optimal gap can be augmented by filling it with a material of high frictional loss, such as mineral wool or glass wool, which can also be advantageous in preventing resonances within the cavity from reducing the effectiveness of the gap.

If we were to build an isolation wall in the way described above, we would by now have a wall and an air gap to provide some isolation, but we would also have the equivalent of an electrical short-circuit between the two walls, created by their common floor, and any contact with the structural walls or ceiling must also be avoided. We must therefore create a structural discontinuity to prevent this bridging of the gap, and the usual method is to 'float' the isolation wall on a resilient material such as rubber, mineral wool, high-density foam, or suitably damped metal springs. In fact, perhaps it will be easier to visualise many of these things if this chapter now bases itself around an actual set of plans. Figure 1.1 shows the plans of a studio built on the ground floor of an apartment building in Granada, Spain, in 1995. The building was of the reinforced-concrete framed type shown in Fig. 1.2, but the situation was complicated by the fact that the bedroom of an apartment was located directly above the area intended for the recording of drums. The isolation structure used the above-mentioned sand-filled concrete block walls and, in this case, the walls, in this instance floated on a high-density synthetic foam of 5 cm thickness.

Float materials must be chosen such that when they are fully loaded by the weight of the structure placed upon them, they are in the middle section of their compression range. If a material is compressed too little, for example by being too hard, it will transmit the vibration. If it is too highly compressed, it will again become hard, or 'solid', for example when a spring is fully compressed, and again the vibrations will be transmitted. Only in the middle loading range will the positive and negative vibrational loads be absorbed by the material. This concept is shown diagrammatically in Fig. 1.3.

1.2.2 Lining the main structure

In the particular case under discussion here, due to the resonant nature of the building shell, the room was lined with 5 cm of high-density mineral wool and two layers of 18 mm plasterboard. The materials were all mounted with a strong, fast drying plaster/adhesive compound (Placoplatre). The plasterboard/mineral wool combination provides a 'mass and spring' combination, or, if the structural wall itself is taken into account, a 'spring sandwich' (mass/spring/mass). In situations like this it is the order of combination which is critical. Sticking the plasterboard to the wall and the mineral wool on top would not be as effective, as the 'spring' (the mineral wool) would no longer be sandwiched between the high mass layers, and waves

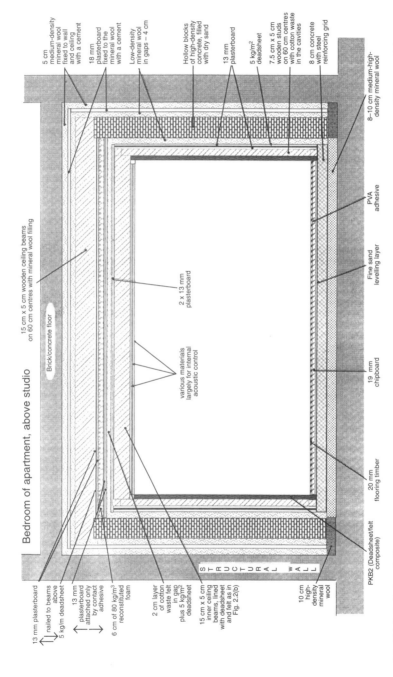

Figure 1.1 Triple isolation shell

(Not to scale)

Bedroom of apartment, above studio

15 cm × 5 cm wooden ceiling beams
on 60 cm centres with mineral wool filling

Brick/concrete floor

5 cm
medium-density
mineral wool
fixed to wall
and ceiling
with a cement

18 mm
plasterboard
fixed to the
mineral wool
with a cement

Low-density
mineral wool
in gaps – 4 cm

Hollow blocks
of high-density
concrete, filled
with dry sand

13 mm
plasterboard

5 kg/m²
deadsheet

7.5 cm × 5 cm
wooden studs
on 60 cm centres
with cotton waste
in the cavities

8 cm concrete
with steel
reinforcing grid

8–10 cm medium-high-
density mineral wool

PVA
adhesive

Fine sand
levelling layer

19 mm chipboard

20 mm
flooring timber

PKB2 (Deadsheet/felt
composite)

10 cm high-
density
mineral
wool

2 × 13 mm
plasterboard

various materials
largely for internal
acoustic control

S
T
R
U
C
T
U
R
A
L

W
A
L
L

13 mm plasterboard
nailed to beams
above

13 mm
plasterboard
attached only
by contact
adhesive

5 kg/m deadsheet

6 cm of 80 kg/m³
reconstituted
foam

2 cm layer
of cotton
waste felt
in gap
plus 5 kg/m²
deadsheet

15 cm × 5 cm
inner ceiling
beams, lined
with deadsheet
and felt as in
Fig. 2.2(b)

Figure 1.2 Typical European building technique

(a) The tube of the concrete pump can be seen extending from the transporting vehicle on the ground, to the top floor of the building where a floor is being cast. The fine steel rods, projecting from the top of the building, will form the core of the next set of vertical pillars

(b) The completed frame of the building can be seen. In the foreground are piles of hollow bricks, which are used to fill in the spaces between the cast-concrete frame. Without the use of damping compounds in the concrete, or the in-filling of the bricks with sand (which would reduce their thermal insulation properties), such structures tend to be very resonant and have poor low frequency sound isolation

(c) A typical interior of a concrete-framed building

penetrating the spring would then impinge directly on the plasterboard. As this would be rigidly coupled to the structural wall, the vibration would then pass directly into the structure. True, the plasterboard would add some mass, but the extra mass added in comparison to the mass of the wall would be minimal. When the mineral wool is in the middle of the sandwich, the sound waves must expend energy in moving the heavy plasterboard, which is damped by a great surface of resilient mineral wool. The mineral wool absorbs much of the vibration, and, as it is fibrous in nature, transmits it only poorly to the heavy mass of the structural wall. As we shall see later, low-density and non-rigid materials have trouble transmitting energy to high-density materials. The mineral wool and plasterboard, bonded to this structural wall, also help to damp the resonances of the structure itself, and so also contribute to the rapidity with which vibrations will decay within the main walls of the building.

Because plasterboard is composed of fine particles, there are great frictional losses as waves try to propagate through it. The acoustic energy is largely turned into heat energy as the particles rub together. Similar frictional losses exist between the fibres of the mineral wool in the sandwich, but here, the bending of the fibres, as they act as springs, causes further losses of acoustic energy. (The complex nature of the losses in fibrous materials are discussed further in Chapter 7.) In such sandwiches as being discussed here, the low frequencies have little problem in passing through the fibres of the

Figure 1.3 Float materials need to be in the centre of their compression range for most effective isolation

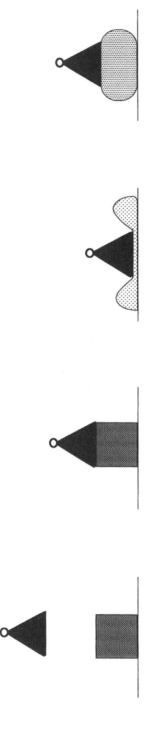

(a) Here we see a weight, suspended above a block of material, intended for isolation

(b) In this illustration, the weight is resting on the block, but the block shows no sign of compression. Its stiffness must therefore be very great, effectively making a rigid coupling between the weight and the ground. Vibrations in the weight (if, for example, it was a vibrating machine) would be transmitted to the ground

(c) Here the isolation material is too soft, and the weight has compressed it down to a very thin layer. The density and stiffness of the material below the weight will thus increase, and the effect will be an almost rigid coupling with the same effect as in (b)

(d) In this illustration, the weight can be seen to have compressed the isolation material to about half its original thickness. The system is at rest due to the equilibrium being found between the gravitational down-force on the weight, and the elasticity of the material. Vibrations in the weight will be resisted by gravity, in the upward direction, and downwards by the mass of the floor and the elasticity of the isolating material. Much of the vibrational energy will be turned into heat by the internal losses in the isolation material, and hence will not be transmitted into the ground

mineral wool, so its only real use in terms of low frequency isolation is when it is used as a spring between two relatively high mass layers. This is indeed the case when mineral wool is used as a 'float' under concrete walls, or floors, again as we shall see later. In fact, fibrous materials *can* be used to stop low frequencies, but usually only when unrealistic thicknesses are used, or when they are located at either ¼ or ¾ wavelength distances from reflective walls. Again, this will be discussed in more detail in Chapter 7.

In Fig. 1.1 it can be seen that there is a further fibrous layer in the 'air gap' between the plasterboard and the floated concrete-block wall. This is lower-density mineral wool, which is used to suppress any longer wavelength sounds circulating in the space. Low frequency waves crossing the gap directly will be little affected by the mineral wool, but situations can develop where waves propagate laterally, effectively circulating in the void. Anything circulating parallel to the walls will have to pass constantly through the mineral wool infill, which will increase the losses and thus help to absorb this energy.

1.2.3 A floated floor

Once we have four isolated walls, we then need a floor. In the case shown in Fig. 1.1, a layer of 8 cm, high-density mineral wool was laid on the floor, on top of a membrane of heavy-duty PVC. The membrane acts as a barrier against any dampness that may exist in the structural floor. A further layer of PVC was then laid over the mineral wool, both to prevent any excess moisture entering from above when an 8–10 cm layer of concrete was poured on top of it, and also to prevent the concrete, itself, from penetrating the upper part of the mineral wool. Once dry, any cement in the mineral wool could make the upper section more rigid, and so would effectively reduce the thickness of the 'spring' layer. Inside the concrete was laid an iron-grid structure for extra strength and the prevention of cracking under load. Once the concrete was thoroughly dry, it was covered with a 2–3 cm layer of dry sand, partly for levelling the surface and partly for acoustic damping of the chipboard, plywood and wood-strip layers which would subsequently form the final surface of the floor of the room. The three wooden surface layers were laid in an overlapping pattern, such that the joints between the boards did not coincide. After each layer had been glued to the other as it was laid, and pinned with small nails before the glue dried, the overall result was a single composite layer, free of any noticeable resonances.

It should be noted here that the whole floor was laid *inside* the isolation walls. There was a space of about 2 cm all around between the floor and the wall, filled with mineral wool. There were two main reasons for this. Firstly, any vibrations transmitted directly into the floor, from bass guitar amplifiers or drum kits, for example, would not be directly coupled to the isolation walls. These isolation walls would therefore only have to deal with airborne noise, making their job much easier. Secondly, if the isolation walls were built directly on the floated floor, the float material under the concrete would have to carry enormous loads around its perimeter, where it would also be carrying the weight of the walls and ceiling. The uneven weight distribution over the surface of the floor would put great stress on the concrete slab, and make a difficult job of having to grade the floating material to be gradually more dense towards the edges. Remember, the float material must be in the middle section of its compression range, as if it is too highly compressed it will not isolate effectively. When the walls and floor are floated separately, the appropriate thickness and density of float material can be chosen for each one. This is explained diagrammatically in Fig. 7.3.

In the case of the room under discussion here, the mineral wool slabs were laid in their normal sense, fibres running horizontally, but it has been shown that greater isolation, for any given thickness, can be achieved by cutting thick blocks of mineral wool across the grain of the fibres, and laying them in smaller sections with the fibres running vertically. In effect it is like laying a floor on a giant scrubbing brush. However, the mineral wool laid in this way needs to be tightly laterally compressed, and restrained around its perimeter, if it is not going to disintegrate as time passes.

1.2.4 Capping the box

All that we now need to complete our isolation shell is a ceiling. A reinforced-concrete ceiling would be an excellent starting point, but when there is little space above, pouring such a structure is difficult. It also takes time to prepare, time to set, and time to dry. Figure 1.1 shows a composite ceiling structure which is relatively easy to construct when there is only minimal space above, and which requires no special skills to do so. The walls in this instance were built to a height of approximately 25 cm below the plasterboard/mineral wool lining of the structural ceiling. The narrowest dimension of the room was only 6 m, so the ceiling could be spanned by timber beams of 20 cm × 7 cm, spaced at 60 cm intervals. In a larger room, however, composite beams of timber and plywood, or steel joists, would be needed to securely span the greater distance. In the latter case, if necessary, timber could be attached to the steel for easier fixing of the subsequent layers.

From Fig. 1.1 it can be seen that two layers of plasterboard, each of 13 mm, were attached to the underside of the joists. This was done with large screws and large washers to distribute the load. The plasterboard layers were fitted with overlapping joints, with the board junctions of one layer connected to one joist, and the joints in the next layer connected to the next joist,

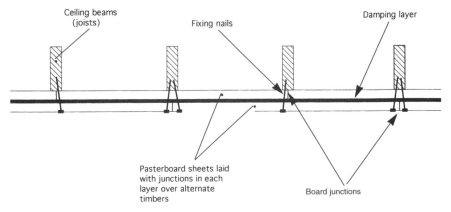

Figure 1.4 Fixing arrangement of plasterboard sheets

and so forth alternately, as shown in Fig. 1.4. When each pair of joists was covered by the first layer of plasterboard, an infill of fibrous material was laid in the spaces between the joists. This stops any 'chatter' in the 20 cm (or so) cavity between the two plasterboard surfaces, formed by the inner layer of the structural ceiling and the upper layer of the isolation shell ceiling, and damps any cavity resonances which could reduce the isolation. In practice, the fibrous material could be mineral wool, glass-fibre wool, or heavy cotton-waste felt.

It will also be noted from Fig. 1.1 that there exists a layer of deadsheet between the two layers of plasterboard. In actual fact it was a layer of bituminous composite sheeting weighing 4 kg/m². Further effectiveness could have been gained by using products such as Revac, or Noisetec LA5 or LA10. These deadsheets have available weights of 5 and 10 kg/m² and are very effective in improving the low frequency isolation. They are plasticised, mineral-loaded products which come in rolls of about 5 m × 1 m and are relatively resistant to fire. Not only do they act as very non-resonant extra mass layers, but they also act as 'constrained layers' between the plasterboards.

Constrained layer technology is very effective at vibration damping. It is shown diagrammatically in Fig. 1.5, where it can be seen that a viscous layer, sandwiched exactly halfway between two identical flexing layers, will try to shear throughout its entire surface as the composite structure is flexed. Shearing forces over such a great area are very strongly resisted, and thus the flexing vibrations are very heavily and rapidly damped. This sort of composite layer can produce around 30 dB of isolation at 50 Hz, which given its simplicity and flexibility can be a very useful option to have available. However, one needs to be careful with the choice of the density of the constrained layer; if it is *too* heavy it will reflect more energy back into the room, but if it is too light it will not absorb sufficiently. A density of 4 kg/m² is a reasonable compromise for such a layer between two 13 mm plasterboards for maximum LF absorption relative to transmission and reflexion. As we pass from the isolation wall to the innermost surfaces of the room though, we need to become gradually more concerned with internal acoustic control, and less with pure isolation, so we will see later how these materials may be used in other ways.

1.2.5 The need for a second ceiling

In the actual building referred to here, the basic structure yielded 30 dB of isolation. A bass guitar and drum kit played on the ground floor was registering peaks in the room of above 113 dB. Measurements taken simultaneously in the bedroom above gave readings of 83 dB, and at such levels it would have been impossible to enjoy a television programme, let alone to sleep. The 30 dB, or thereabouts, of extra isolation provided by the floated room and composite ceiling described in the last few paragraphs could reduce that to about 55–60 dB (bass guitar and bass drum fundamentals being well below 50 Hz, see previous paragraph) but 60 dB is the level of a normal conversation in the room, so isolation would still be far from adequate. Given the relatively massive structure of the floated walls and floor, it is clearly the ceiling, in this case, which is the weakest surface of the room, and unfortunately it is the surface directly facing the problem of the

Figure 1.5 Constrained layer damping principle

(a) Damping material fixed to flexible panel

(b) When panel is flexed in this direction, the damping material (shaded layer in diagram) will stretch

(c) When flexed in this direction, the damping layer will compress

(d) Damping layer sandwiched between two similar flexible panels

(e) When flexed, the top layer will stretch and the bottom layer will compress. The damping layer, however, will remain the same length, but, if viscous, will attempt to shear throughout its entire area, down its centre line. The forces resisting this are very great indeed and, consequently, so is the damping provided by the constrained layer

bedroom above. To increase the isolation, it was decided to add a very different type of layer below the first composite ceiling. This consisted of a layer of reconstituted polyurethane foam, with a thickness of 4 cm and a density of 80 kg/m^3. It was attached to the plasterboard above by contact adhesive, and below were glued two further layers of 13 mm plasterboard with their joints overlapping so that no edges of the two layers were coincident.

It often seems to worry people when they see such a weight of plasterboard suspended solely by adhesives from a foam ceiling. The foam is similar to that frequently found in flight cases – the multi-coloured stuff – but it is cut in a different orientation to the usual manufacturing process. I often demonstrate with a small square, say 10 cm × 10 cm, which I stick to a ceiling, then continue to attach two 10 cm squares of plasterboard to the foam using the same contact adhesive. 'Yes' people often say, 'but that is only a small piece, you are planning to put tons up there!' What they so often fail to realise is that the kg/m^2 pull is exactly the same, whether it is 10 cm^2 or 100 m^2. The foam itself has a resistance to traction of around four tons per square metre, and the strength of the contact adhesive used is only slightly less, which is over one hundred times the strength necessary to support a double 13 mm layer of plasterboard.

This sort of structure acts somewhat like a boxer's punch-bag, with the plasterboard taking the initial energy of the blow, and the foam subsequently absorbing it as it is trapped between the two mass layers. A deadsheet layer was not used between the plasterboard sheets, in this instance, as the plasticised sheets are apt to reject the adhesive and weaken the structure. The whole ceiling thus formed a highly damped vibrating membrane, perhaps with a total isolation at 40 Hz of around 40 to 50 dB.

The reason why different types of isolation structures were used for this ceiling, as opposed to just using more of the first technique, is that all isolation systems have certain characteristic resonances. These resonances can cause weak spots in the isolation, so if a second identical system is used, these weak spots will occur at the same frequency. However, when an entirely different isolation system is used, it is less likely that any resonances will coincide so, effectively, each system largely covers any weaknesses in the other.

1.2.6 First considerations of internal acoustics

It was not, however, solely for its isolation properties that this type of ceiling was used. As will be described in later chapters, the isolation shell can have its effects on the internal acoustics of the finished room. Although the concrete wall and floor structures are good at preventing sound *transmission*, they do so by reflecting much of the sound back into the room. Effectively, they are containment shells. By contrast, using a highly resilient ceiling provides isolation by achieving a great deal of low frequency *absorption*, which, in conjunction with the walls and ceiling, achieves perhaps 50 dB of wideband isolation without an undue low frequency build-up in the space within, though they do reflect mid and high frequencies.

1.2.7 General comments on the shell

Thus, what we now have is a fully floated isolation shell, penetrated only by doors, ventilation ducts and possibly windows, and ready for the internal acoustic treatment. As mentioned earlier, this is not intended to be a book on sound isolation, which is a truly enormous subject, but an isolation shell should normally be a starting point for most acoustic designs for the studio rooms. Without at least a little understanding of the point from where we usually begin the internal designs, many of the concepts of the following acoustic designs would be without foundation, and it would leave too many gaps in the information to which we may need to refer back. At least a basic understanding of some of the isolation concepts is therefore a necessary starting point.

The isolation shell described here is not in any way definitive, but it is an interesting example as it incorporates a number of different techniques, a description of which helps to give a reasonably broad grounding in the different processes which we may encounter. There is a huge variety of products on the international market for isolation materials, and different designers will have many of their own favourite materials and methods. There are literally dozens of ways of approaching the design of a room similar to the one described here, and costs, availability of materials, availability of appropriate labour, weight considerations, humidity or insect problems, and many, many other things may all have a bearing on the precise method chosen.

In the specific case described, tight budget restraints, availability of local labour and materials, and speed of construction all had their parts to play in why the room was built the way it was. In the end, after the construction of an internal acoustic control shell and surface treatments, it eventually yielded 83 dB of isolation from a bass guitar and drum kit in the studio to the bedroom directly above. Above 100 Hz, the isolation increased rapidly, and so for acoustic instrument use, except for bass drums and tympani, even in the middle of the night it would be impossible to play at such a level as could be detected above the minimum background noise in the bedroom above. The 113 dBA of the typical bass and drums set-up produced a noise level of 30 dBA in the bedroom. It was considered that, whilst some 'extreme' cases could perhaps produce 120 dB in the studio, the owners of the studio in Granada considered that these cases would be too few and far between to render necessary the expense of extra isolation. Indeed, after one year of unrestricted use, they had not received one, single complaint from their neighbours.

There also had been a battle between the space needed for isolation and the demands of the owners for the maximum available internal space. To yield 90 dB of isolation it would have been possible to use 10 cm foam on the ceiling, an extra 2 cm of concrete on the floor and, perhaps, the use of a 10 kg/m^2 deadsheet in the isolation sandwich of the ceiling, plus a little more mineral wool of higher density between the joists. However, these measures could have reduced the available ceiling height by around 10 cm, and as there was only 3 m 44 cm of available height in the original building, the owners absolutely refused to allow another 10 cm to be 'lost' to isolation. The results were ultimately deemed acceptable by the studio owners, the musicians and the neighbours above. Compromise is very much a part of the job of a studio designer; however, I have learned the hard way that when the compromise point seems *too* limited on any aspect of the design, it is better to refuse the job, no matter how much it may be wanted by *any* of the parties involved. To go ahead tends only to lead to bad feeling and future problems. Some spaces are just not suitable to be studios, and that point must be clearly understood. In the excitement of the planning stages, many studio owners agree to performance limitations which the designer may advise them against, but they often rapidly forget their acceptance of the limitations once they have spent all of their money and the reality of the limitations strikes home. Clearly, warning of expected limitations is an important part of the work of a studio designer, and those warnings will largely be based on a wealth of experience and they should be very carefully considered. If this type of advice is dismissed too casually, it can prove very expensive in the long term.

Anyhow, what has so far been described in this chapter is a viable isolation shell, which in the vast majority of cases will consist of four vertical walls, a floor, and a ceiling. The shell, at this stage, will probably sound highly unmusical as it will possess strong resonances at frequencies which correspond in wavelength to the room dimensions. The reason for this is because most of the isolation has been achieved by reflecting the energy back into the room, so clearly, something must be done to rectify this situation. The following five chapters will look at what we can do inside the isolation shells in order to create environments which are sonically appropriate for various types of musical performances and recordings. However, before we proceed any further, perhaps we should consider what goes on inside the bare isolation shells, as this will give us some insight into why the work described in the following chapters is necessary.

1.3 Modes and resonances

Sound consists of tiny local changes in air density which propagate through the air as a wave motion at the speed of sound. The speed of sound is around 340 m per second at normal room temperature and, although it is temperature dependent, it is independent of variations in the ambient pressure, and is the same at all frequencies. The frequency of a sound wave is measured in cycles per second (c/s or cps) known more usually nowadays as hertz (Hz), and is usually represented by the symbol f. The distance that a sound wave travels in one cycle at any frequency is known as the wavelength, represented by the symbol λ (lambda), and is measured in metres. The speed of sound is represented by the symbol c. The relationship between wavelength, frequency and the speed of sound is simple; wavelength is equal to the speed of sound divided by the frequency; or $\lambda = c/f$. So, for example, a sound wave at a frequency of 34 Hz has a wavelength of 340/34 = 10 m.

As a sound wave propagates away from a source in a room, it will expand until it reaches a reflective room boundary, such as a wall, from which it will reflect back into the room. The reflected wave will continue to propagate until it reaches other boundaries from which it will also reflect. If there is nothing in either the room or the boundaries to absorb energy from the wave, the propagation and reflexion will continue indefinitely, but in practice some

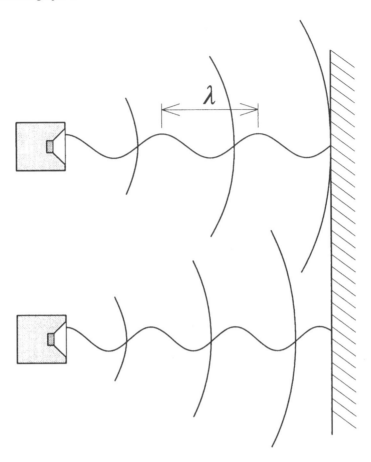

Figure 1.6 Phase relationship of reflected waves. Sound waves of different frequencies, and hence of different wavelength, emanating from a sound source with the *same* phase relationships, will, after travelling identical distances to a reflective boundary, arrive at the boundary with *different* phase relationships. From the above figure, it can be seen that the two waves leave their respective sound sources at a positive peak of their cycles, but arrive at the wall quite differently

absorption is always present and the wave will decay with increasing time. The point on the cycle of a sound wave (the *phase* of the wave) when it reaches the boundary depends upon the distance to the boundary and the frequency of the wave. Figure 1.6 shows how waves at different frequencies, propagating from the same source, arrive at a boundary with different phases.

A rigid boundary will change the direction of propagation of an incident sound wave, but will maintain its phase, so the phase of a reflected wave can be calculated from the *total* distance propagated from the source. If this total distance is equal to a whole number of wavelengths then the wave will have the same phase it started with. When two boundaries are parallel to each other, a sound wave will reflect from one boundary towards the other, and then reflect back again to where it started, continuing back and forth until its energy is dissipated. If the distance between the boundaries is such that the 'round trip' from the source to the first boundary, on to the second boundary, and back to the source is a whole number of wavelengths, then the returning wave will have the same phase as the outgoing wave and will serve to reinforce it. This situation is known as resonance. Resonances can also occur due to reflexions from multiple boundaries; the necessary requirement being that the sound wave eventually returns to a point with the same phase as when it left. One can imagine a whole set of possible combinations of reflexions in a typical room which allow the wave to return to its starting point, and therefore a whole set of frequencies for which resonance will occur. In fact, in theory, every room has an infinite number of possible resonances.

As stated above, if there is nothing in the room or the boundaries to absorb energy from a sound wave, a short duration sound pulse (or transient) emitted from a source will propagate around the room indefinitely. Of the infinite number of possible paths that the wave can take, only those which correspond to resonances at frequencies contained in the pulse will be continually reinforced; all other paths will decay rapidly. After a short time, the resulting sound field can be thought of as simply a sum of all of the resonances that have been excited. These resonant paths are known as the natural modes of the room and the resonant frequencies are known as the natural frequencies, or 'eigentones', of the room; both are determined uniquely by the room geometry ('eigen' is German for 'own', so the eigentones are a room's own particular, natural, resonance frequencies).

When sound absorption occurs within the room or boundaries, resonant modes still exist, but the wave will decay at a rate determined by the amount of absorption. To maintain a given sound level in a room in the presence of absorption, the source needs to be operated continuously at a level dependent both upon whether or not resonant modes are being excited, and the amount of absorption present. When the sound source emits a transient signal in the presence of absorption (for example, switching off a continuous signal), many different paths – not just resonant modes – will be excited, but after a short time, only the resonant modes will remain; the room will 'ring' at the resonance frequencies until the modes decay. The reverberation time of a room is a measure of the average rate of decay of the sound in the room when an otherwise continuous sound source is switched off; it is the time taken for the sound level to fall to -60 dB relative to its initial, continuously excited, level. As the amount of absorption is increased, the sound level at the resonant frequencies will reduce but the bandwidth of each mode (the range of frequencies over which the mode can be excited to a significant degree) will increase. When the boundaries are fully absorbent the room modes no longer exist (an anechoic chamber).

When sounds such as speech or music are heard in a room, the level of the continuous components of the sound will be determined by whether or not they coincide with any room resonances that are excited. The transient components will 'hang on' at the resonance frequencies after the transient has gone.

The above topic will be expanded upon as we progress, but our job, in whatever direction we seek to develop the acoustics of a room, will be to control the room modes. In such a way, we can create rooms which have the required degree of musicality for our purposes, such that the response is what we want, and is not restricted to a response which is dominated by the eigenfrequencies of the isolation shell. In order to achieve this, we can use geometry to control the pathways (and hence the resonance frequencies), and absorbent materials to both control the level of reflexions and the spread of frequencies which contribute energy to specific modes.

Chapter 2

Neutral rooms

Historically, recording studios, and by the word 'studios' I mean the recording spaces as opposed to the control rooms, have been relatively neutral environments. This has been partly due to past recording studios having to cater for a wide range of recordings. Too much bias towards the needs of one specific type of music could lead to a restriction in the amount of work available for a studio, and, furthermore, it was generally held that the responsibility for the production of the sound to be recorded was within the domain of the musicians. The function of the recording studio was seen as being to record the sounds which the musicians produced, as faithfully as possible. The days of a somewhat more creative side to the recording process had not arrived.

However, contrary to what may often be expected, the 'true' sound of a musical instrument (that is, the sound as its designer intended it to be heard) is not that which it would make in an anechoic chamber. This is because instruments were developed in the circumstances of more reflective or reverberant surroundings, and it is frequently the combined direct *and* reflected sounds which constitute the 'true' sound, that is, the sound as its designer *intended* it to be heard. Concert halls have long been rated by musicians according to how well they can perform in those halls. Players of acoustic instruments need a feedback from the performing space, as frequently the sound emissions from the instruments are inadequate to give, directly to the musicians, the sensations required for optimum performance. String sections need to hear string *sections*, not a group of individual instruments. When they hear a section, they play as a section; when they hear separate instruments, their playing also often fails to gel into a single, homogeneous performance

Flautists seem almost always to need some reverberant help for their playing, be it natural, or fed electronically into their headphones. Woodwind players also seem to dislike too dry an ambience in their performing space. An anechoic chamber is a truly awful environment in which to play any instrument, and such an environment is not going to inspire any musician to the heights of creativity. Creativity is supremely important, as certainly as far as I am concerned, an uninspired performance is hardly worth recording.

So, if we do not mean a clinically accurate recording space when we speak of 'neutral' spaces, then what *are* we talking about? Essentially, a neutral environment is one which provides sufficient life to allow enough of the character of an instrument to be realised, but which does not overpower the instrument with the character of the room itself. This means a smoothly sloping reverberation time (or decay time) curve, together with discrete reflexions which add life, but do not dominate the natural sound of the instrument. Normally, such rooms will have reverberation times which rise as the frequency lowers. This is a function of most enclosed spaces, other than very small ones, and when one considers the fact that most instruments have been developed for performance *in* such spaces, then a recording area with similar characteristics would not be deemed unnatural. Rooms of different sizes, shapes and structures will have their own characteristic acoustics, but as long as those characteristics do not add any significant timbral change to the instrument, they can be considered to be neutral.

In general, it is easier to make large neutral rooms than small ones. This is because of two main reasons. Firstly, in large spaces, the resonant modes which exist tend to be more evenly spaced across the frequency spectrum, whereas in small rooms, especially at lower frequencies, they tend to cluster together. These clusters, particularly in the upper bass region, can become strongly audible due to the concentration of energy which often begins to add a strong character to the sound of the instruments in the room. Secondly, in larger rooms, there is a greater period of time between the emission of sound from the instruments and the arrival of the reflexions. True, floor reflexions are returned at similar time intervals in all sizes of rooms, but these are usually relatively innocuous single reflexions, free of resonant characteristics. The more noticeable resonant modal energy must exist between at least two surfaces, and hard, parallel floor/ceiling combinations are almost always avoided in studios. In large rooms this greater period of time before the first reflected energy returns to the instrument allows more time for the direct sound of the instrument to stand alone, and thus establish itself clearly in the perception of the listeners.

A further two reasons also lead to the later reflexions having less colouring effect. Reflexions from greater distances have further to travel, so when they do return, they will do so with generally less energy than those travelling back from shorter distances (given the same reflecting surfaces). What is more, when reflexions arrive much more than 40 milliseconds (ms) after the initial sound, they tend to be perceived by the brain *as* reflexions, whereas those arriving before a time of 40 ms has elapsed will almost certainly be heard as a timbral colouration of the instrument, and will not be perceived as discrete reflexions. In a large room, therefore, the resonant modes and reflexions are heard as separate entities to the direct sound of the instruments, and unless the direct sound of an instrument is swamped by room sounds which are unduly long in time, or high in level, its natural characteristic timbre will be clearly heard.

Generally, a recording room can be considered to be in the 'acoustically small' category if it is impossible to be less than 4 m or 5 m from the nearest wall surface. Judicious angling of the ceiling together with careful use of absorption and diffusion can permit the use of ceilings of 4 m or less with relative freedom from colouration, so an acoustically small room, for recording purposes, would tend to be less than 10 m × 10 m × 4 m.

2.1 Large neutral rooms

To build an acoustically large neutral room is not a particularly difficult exercise as long as a few basic rules are followed. Parallel hard surfaced walls should be avoided, as these can support the strong axial modes which develop, reflecting backwards and forwards between the parallel surfaces. The effects of such reflexions on transient signals are those of 'slap-back' echos, or series of repeats, usually with pronounced tonal contents, which are often quite unmusical and unpleasant in nature. The parallel surface avoidance rule also relates to floors and ceilings. However, we should make a distinction here between what is actually heard in the room, and what is heard via the microphones. Although the *ear* is less troubled by vertical reflexions, as it is much more sensitive in the horizontal plane, most microphones make no distinction between horizontal or vertical reflexions. Therefore, a floor/ceiling problem or a similar wall/wall problem will be detected by most microphones in exactly the same way. Even though the ear may hear them very differently when listening directly, the perception when listening to a recording will be the effect as detected by the microphones. Consequently, when considering the subjective neutrality of a room, we must consider it from both points of reference: the human ears, *and* the microphones.

Unless two rooms are absolutely identical, not only in shape, size and surface treatments, but also in the structure of their outer shells, they will not sound identical. In reality, there are thousands of 'neutral' recording spaces in the world, but it is unlikely that any two will sound identical. The achievement of acoustic neutrality is all a question of balances and compromises, but unlike the neutrality needed in control rooms, where repeatable and 'standard' reference conditions are needed, such uniformity of neutrality is not required in the studios. The concept of neutrality in a recording room therefore occupies a region within an upper and lower decay limit of what a room may add. All that is required is that the room sound is evenly distributed in frequency and is subservient to the sound of the instrument(s).

A parallel exists in the realm of amplifiers. Guitar amplifiers and hi-fi amplifiers are quite distinct breeds, and normally cannot be interchanged. Guitar amplifiers have relatively high levels of distortions, but those distortions are chosen to be constructive and enhancing in terms of the sound of electric guitars. However, recorded music played through a guitar amplifier and loudspeaker will sound coloured, and will suffer from a lack of definition. The result will certainly not be hi-fi. Conversely, a guitar played through a hi-fi amplifier and loudspeaker (the neutrality of which is more akin to control room neutrality) will be unlikely to sound full bodied or powerful. No matter how loud it is played it will sound weak, until eventually a level is reached where the sudden onset of gross harmonic distortions will create unpleasant, disonant sounds. In a similar way, it is thus entirely justifiable for a neutral recording room to add to the character of an instrument played within that room, as long as that character enhances and supports the instrument and does not in any way become predominant itself. Anything, therefore, in the sound *production* side of the record/reproduce chain, can be considered to be an extension of the instrument, and hence subjective enhancement is usually desirable. Conversely, things in the *reproduction* or quality control sides of the chain must be transparently neutral, in order to allow the production to be heard as it was intended to be heard.

2.2 Practical realisation of a neutral room

Neutral rooms are desirable for many types of acoustic recordings, but modern thinking places much more emphasis on the comfort of the musicians than was encountered in a great majority of the coldly neutral rooms which were prevalent in previous years. At each stage of our design, therefore, we need to consider its effect on the musicians, as well as on the purely acoustical requirements. That being said, it seems to be very widely accepted that floor reflexions are in almost all cases desirable, and indeed most live performance spaces have hard floors, so let us begin the design of our large neutral room with the installation of a hard floor.

2.2.1 Floors

Hard floors can be made from many materials, but they generally subdivide into vegetable or mineral origins. On the vegetable side we have a great variety of wood-based choices. There are hardwoods, softwoods, wood composites such as plywood, veneered MDF (medium-density fibreboard), fibreboards, parquet, and reconstituted boards to name a few of the most common types. In the mineral domain we have stone in its various forms, ceramic tiles, and concrete with resinous overlay, which is often used in television studios where cameras must roll over an extremely smooth, unjointed surface. Usually, in large rooms, wood prevails. It is aesthetically warmer, thermally warmer, less prone to slippage, both by people and instruments, and it is generally richer acoustically. Instruments such as cellos and contrabasses rely on the floor contact to give them a greater area of soundboard as their vibrations travel through the wooden floor. Mineral based floors do not 'speak' in the same way. Wood also represents more closely the flooring which musicians are most likely to encounter during their live performances, and, where possible, studios should seek to make the musicians feel at home. The necessity for this cannot be over-emphasised. The exact nature of the floor structure will usually depend on a multitude of factors concerning the structure and location of the building, but these things will be dealt with in later chapters.

2.2.2 Shapes, sizes and modes

Let us consider that we are starting with a concrete shell, already isolated acoustically from the main building structure. Our worst case starting point would be to have a cubic room, with all dimensions (length, breadth and height) equal. As we saw in Chapter 1, when a sound radiates from a point in a reflective room it will expand at the local speed of sound, and rebound from surface to surface until all its energy is dissipated as heat, due to losses in the air and in the surfaces of the walls on each contact. Some of these reflexions repeatedly travel backwards and forwards along the same paths, and become resonant modes. There are three basic types of resonant modes which build up and tend to reinforce themselves. *Axial* modes exist between two parallel surfaces, travelling parallel to the other four surfaces of any six-surfaced room (four walls, ceiling and floor). *Tangential* modes travel around four surfaces, and remain parallel to the remaining two. *Oblique* modes travel round all six room surfaces, and are parallel to none.

In the case of a cubic room, where all the pairs of parallel surfaces are equally spaced apart, the axial modes will all be similar in pathlength, and hence will all have similar resonances. This will lead to a strong resonant build-up of energy at those frequencies. Furthermore, the axial modes are the ones which are considered to contain the most energy, and so those frequencies whose wavelengths correspond with the dimensions of the room will strongly predominate, giving the room a highly tuned, strongly resonant character. Such a room would be a 'one note' room, with overpowering resonance destroying the musicality of any instruments played in the room. At the other extreme, a room of rectangular plan with dimensions of height, breadth and length in the approximate proportions 1:1.6:2.4 would render the most varied assortment of modal frequencies, and hence the overall least coloured sound. It was long held that this type of room should form the basis of 'standard' listening rooms for the assessment of domestic equipment, but it has more recently been pointed out that the aforementioned modal properties only relate to an empty room. As soon as one installs equipment, people, decorative surfaces and so on, the smoothness of modal distribution is lost. Nonetheless, a room of such proportions would be a *much* better starting point than the cubic room, although these proportions only hold good for rooms that are neither extremely small nor the size of large concert halls.

Irregularly shaped rooms tend to create a greater modal resonance spread, as it becomes difficult for the sound waves to 'find' paths of equal length on subsequent reflexions. The modal resonanaces tend to be of the tangential or oblique forms, which also generally contain less energy than the axial modes, and become lower in their 'Q' (or tuning) as the energy becomes spread more broadly, being less tuned to specific notes. The natural reverberance of these rooms is usually smoother, with fewer predominating frequencies. In all the above cases, however, the most persistent problem is how to control the more widely spaced modes in the lowest octaves of the audible range, where wavelengths are long, even when compared to any possible angling of the walls.

So far, in the case of the isolation shell, we have been looking at a rather bad (musically speaking) live room, although in virtually all practical cases we do actually begin with live shells because few suitable construction materials for an isolation shell are acoustically very absorbent. What is more, the shells may be given to us as a fixed starting point, for example when a company already has a building which they insist on using as a studio. In such cases, it is the job of the designers to get the best out of what is presented to them. It is perhaps less common for studio designers to be given a space large enough to create any desired shell shape inside, and less common still to have the opportunity of designing a building from scratch. Such are the realities.

2.2.3 From isolation shell towards neutrality

Perhaps, then, in our quest for neutrality, it would be wise to look at a relatively difficult, but nonetheless likely, case of a shell of 15 m × 10 m × 5 m high. It is rather awkward because the length and breadth are exact multiples of the height, so resonant modal frequencies of the floor/ceiling dimension can be supported two and three times over in the breadth and length. Strong irregularities at the resonant frequencies (34 Hz and 68 Hz for the breadth and length, respectively) would exist in different locations in the room, depending on whether the sound sources *or* the microphones were located at nodes or anti-nodes, where the pressure of the cross-modes were at a minimum or maximum. The resonance at approximately 23 Hz, being the-

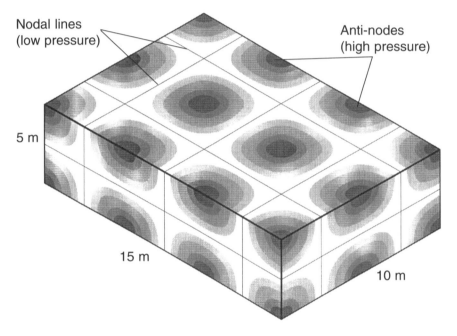

Nodal lines
(low pressure)

Anti-nodes
(high pressure)

5 m

15 m

10 m

Figure 2.1 Modal pattern, at resonance, in a room where the length and breadth are exact multiples of the height

first mode of the length dimension, would perhaps be less troublesome; firstly because it is so low, and secondly because it cannot be supported in the other dimensions of the room. As can be seen from Fig. 2.1, the pressure distribution around the room would be most uneven, and strongly differentiated from one point in the room to another.

There is no simple surface treatment which can effectively stop the resonant modes of long wavelength. In fact they are remarkably resilient, especially so in the type of isolation shell which we would probably be faced with here. Remember, that when sound strikes a wall, there are three possibilities. It can be transmitted *through* the wall, it can be absorbed *by* the wall, or it can be reflected *from* the wall. Obviously, any significant transmission *through* the wall would not be providing us with the required sound isolation for a studio. Absorption is not likely to exist to any significant degree in the materials of an isolation wall of typical, heavy construction, so isolation is likely to be achieved mainly by reflexion; containing the sound inside the room until it eventually dissipates as heat. Eventually, decay it will; but perhaps not until it has bounced around a couple of hundred wall surfaces.

Once a sound has left its acoustic near-field, which is around one wavelength of its lowest frequency, it will begin to decay by 6 dB for each doubling of the distance away from the source. This is the loss due to the expansion of the wave, and is why sounds grow quieter as we move away from their sources. As the sound wave expands, its power is distributed over a greater total area, so the power is quartered for each doubling of the diameter of the expansion sphere. (The surface area of a sphere, over which the energy is distributed, quadruples as the diameter doubles.) In a reflective room, the energy is constrained, and the natural decay is superimposed with reflexions. The overall sound-field in the room therefore does not decay in the same way as in an unbounded space, because although the path of any single wave still follows the same loss of sound pressure level (SPL) as it would in a boundary-free space (free-field), at least in a perfectly reflective room, it is continually being folded back over the same space within the room. However, a perfectly reflective room does not exist, so as a sound travels around a room, each subsequent reflexion from a wall will rob it of some power, the precise amount being a function of the coefficient of absorption of the walls and the angle of incidence of each strike. The direct sound will rapidly disappear once it has passed the ears of the listener (or microphone) but it will soon be replaced by the enormous number of reflexions which build up to produce the reverberant field, which will be characteristic of the physical properties of the room.

In the bare containment shell (isolation shell) as described in the last chapter, some measures can be taken in its design and construction to control excessive reflexions of low frequencies by incorporating large areas of low frequency absorption. Low frequencies can only be effectively controlled by physically large absorbers, which, in the case of panels, are typically of dimensions at least in the order of half a wavelength of the lowest frequency to be absorbed. It is often very space-consuming to have to produce significant amounts of bass absorption in the internal treatments of the room, so whatever absorption exists in the isolation shell will make the final room design easier. As the low frequencies will largely penetrate the lighter weight acoustic control structure, any absorption in the isolation shell will be effective in aiding the internal control, even though it is outside the 'box' of the final room. Here is one area where sound containment shells and internal acoustics overlap inextricably.

Anyhow, back to our objective of producing a neutral room. So far, we have a 15 m × 10 m × 5 m containment shell and a wooden floor. There will be *some* low frequency absorption due to the ceiling structure, but the room will still possess a reverberation which will be both long, powerful (easily colouring the direct sound) and possessing strong resonances, around 70 and 140 Hz, which relate to the wavelengths with multiples of 5 m. As with so many other aspects of studio design, there are some widely varied solutions to this problem, and designers will each have their favourite methods, so I should perhaps address this problem as though it were a job facing me on my drawing table.

2.3 A practical design approach

My initial approach would be to construct a timber-framed internal box structure. Considering the size of this room, and the need for the walls to support a considerable weight of ceiling, a frame structure of 10 cm × 5 cm softwood vertical studs would be constructed, situated on 60 cm centres. This spacing is sufficient for strength, and is conveniently half of the width of most sheets of plasterboard, which tend to come in sizes of 1 m 20 cm × 2 m 50 cm, 2 m 60 cm or 3 m. This is important, because to produce our low frequency absorption system, the next step is to cover the rear side of the stud wall with the same plasterboard/deadsheet/plasterboard sandwich as used in the ceiling described in Chapter 1. The wall frames are usually built on the floor, horizontally, where the boards and deadsheet can be laid over the frame and conveniently nailed to the wooden frame, where possible with large headed, rough coated, galvanised nails. A further layer of thick cotton-waste felt, or other fibrous material, is usually fixed to the surface before the wall is raised into a vertical position. When the four walls are in place and nailed together at the corners, they are then capable of carrying the weight of several tons of ceiling.

The spacing between this internal wall and the isolation wall is important, as a larger space will usually produce more absorption, especially at low frequencies. However, once again, studio owners usually want to see in the finished results as many as possible of the square metres of floor space that they are paying for. It seems that no amount of trying to convince them that they are *hearing* a superior sound, *produced* by the space that they are paying for, has much effect on many of them. We usually therefore have to use a further quantity of mineral wool, glass wool, or cotton-waste felt type materials in order to provide some augmentation of the absorption.

With the fibrous material applied to the rear of the acoustic control walls, as described above, 5–10 cm of space between the isolation walls and the acoustic walls will usually suffice for the acoustic control of studio rooms. The cotton felt described is about 2 cm thick and of quite high density (40–50 kg/m^2). For safety, it is also treated with a substance to make it self-extinguishing to fire, but in cases where absolute incombustibility is required, mineral-based fibrous materials can be used, though they are somewhat less comfortable to work with. A further one or two layers of the felt are then inserted in the spaces between the studs (the vertical timbers). These are cut to fit quite well in the spaces, and are fixed by two nails at the top of the frame. The felt not only suppresses resonances in the closed cavity, which will be formed when the front surfaces are fitted, but also provides another frictional loss barrier through which the sound must pass twice as it reflects from the outer layers, once in each direction.

If additional isolation and absorption are required, it is possible to add a layer of material, such as Noisetec PKB2, over the felt on the rear of the wall. PKB2 is a kinetic barrier, and is a combination of a cotton-waste felt layer, of about 2 cm thickness, bonded by a heat process to a mineral-loaded deadsheet, the composite weighing about 5 kg/m^2. If this is nailed over the felt, with the deadsheet to the first felt layer, it forms a deadsheet barrier, sandwiched between two layers of felt. Indeed, in the neutral rooms of the type being discussed here, PKB2, or a similar combination, would most likely

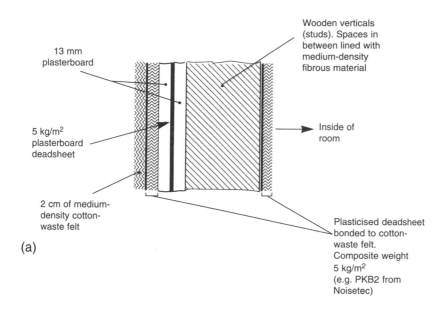

13 mm
plasterboard

Wooden verticals
(studs). Spaces in
between lined with
medium-density
fibrous material

5 kg/m² plasterboard
deadsheet

Inside of
room

2 cm of medium-
density cotton-
waste felt

(a)

Plasticised deadsheet
bonded to cotton-
waste felt.
Composite weight
5 kg/m²
(e.g. PKB2 from
Noisetec)

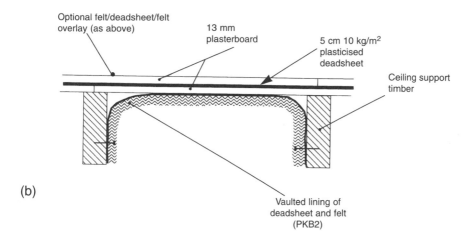

Optional felt/deadsheet/felt
overlay (as above)

13 mm
plasterboard

5 cm 10 kg/m²
plasticised
deadsheet

Ceiling support
timber

(b)

Vaulted lining of
deadsheet and felt
(PKB2)

Figure 2.2 Acoustic control materials: (a) wall; (b) ceiling

form the first of the layers on the *internal* side of the stud walls. With a dead-sheet/felt front, covering a 10 cm-deep cavity, partially filled with fibrous material, and backed by a double sandwich of plasterboard/deadsheet/ plasterboard then felt/deadsheet/felt, followed by a further sealed air cavity before the structural or isolation wall, a very effective low frequency absorption system can be constructed, which will also be absorbent in the higher frequency ranges. This can be done without special tools or skills, and all in a space of about 22 cm. The construction is shown diagrammatically in Fig. 2.2(a).

The reason for the multiple layers of different materials is because different materials and techniques absorb by different means, and are effective in different locations and at different frequencies. Large panel absorbers, made out of plywood, for example, can produce high degrees of absorption, but they will tend to do so at specific frequencies, as they are highly tuned devices. Absorbers are considered to have a Q, or quality factor, dependent upon the degree of tuning. A high Q (highly tuned) absorber, may well absorb strongly at 70 Hz, but may absorb only minimally at 60 and 80 Hz. Obviously, to absorb a wide range of frequencies by such techniques, we would need to have many absorbers, but we may have problems finding sufficient space in the room to site them all.

If we lower the Q of the absorber, by adding damping materials, we will reduce the absorption at the central frequency, but we will widen the frequency range over which the absorber operates. We can therefore achieve a much better distribution of absorption by filling a room with well damped absorbers than by filling it with individual, high Q absorbers, in which case the absorption in any given narrow band of frequencies would be localised in different parts of the room. Another advantage of lower Q absorbers is that the resonances within them decay much more rapidly than in high Q absorbers. Resonators of a highly tuned nature, whilst absorbing much energy rapidly, will tend to ring on after the excitation signal has stopped, and hence may re-radiate sound after an impulsive excitation. Further details of the absorption mechanisms will be dealt with in later chapters, so let us now return to our neutral room.

The ceiling can be dealt with in precisely the same manner as the walls, but, given the 9–10 m minimum space across this room, joists of either steel or plywood sandwiches would seem appropriate. A typical plywood beam cross-section is shown in Fig. 2.3. The only significant difference between the wall and the ceiling structures would be on the inside, as shown in Figure 2.2(b), where the PKB2, or similar material, would be placed in arches, between the joists.

2.3.1 Design considerations

A studio room is an instrument in itself, so what we have by now managed to do is to destroy an instrument, but, if I am faced with a troublesome instrument such as a room of these dimensions might be (15 m × 10 m × 5 m) it is at times prudent to acoustically destroy it, and then re-build it predictably. With troublesome dimensions, the total suppression of the unwanted acoustic characteristics can be difficult to remedy by conventional means, and time and experimentation can be needed to assess the remedial work. For this reason the acoustical destruction of our problematical room is often a wise choice. So, to finally change the characteristic of this room in such a way that its acoustic dimensions no longer represent its physical dimensions in any simple manner, ceiling absorbers could be fitted as in Fig. 2.4, along with a panel absorber along one wall, as in shown Fig. 2.5.

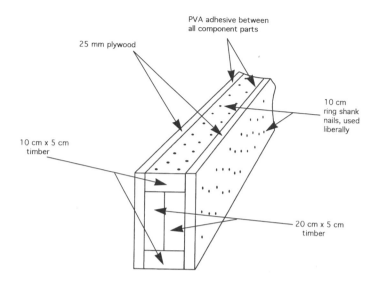

Figure 2.3 Plywood beam construction – a 30 cm × 15 cm beam of immense strength (all component parts glued and nailed to adjacent parts)

By this stage we do not have a neutral room in the recording sense, but, rather, except for the reflective floor, we have something very much approaching an anechoic chamber with a small amount of well-damped low frequency modal energy. Such a room could perhaps be excellent for measurement purposes, or could perhaps form the basis of one of the currently favoured concepts of control room designs, but as a room for recording music it would surely be found to be wanting.

In the case of the studio room which we are discussing here, we need to create an acoustic which enhances the sound of the instruments without unduly announcing its own presence. We need a room which, as far as possible, favours all notes reasonably equally, neither 'wolf' notes, which stand out due to their coincidence with room resonances; nor causing other notes to have to be 'forced' to fight their suppression. The room should have a sonic ambience in which as wide a range of musicians as possible feel comfortable, both in themselves and with their instruments. Such a room allows a wide range of choice for the recording engineers in the positioning of microphones. It also allows a great deal of freedom in the positioning of the different musicians, either for the purposes of improved eye-to-eye contact (which can be very important to them) or for purposes of acoustic separation. However, the wall and ceiling absorbers which we have proposed thus far to overcome the room problems would be rather too absorbent for our requirements of 'neutrality'. So, after controlling the room, we will have to selectively brighten it up, by means that we shall explain shortly. Before doing so,

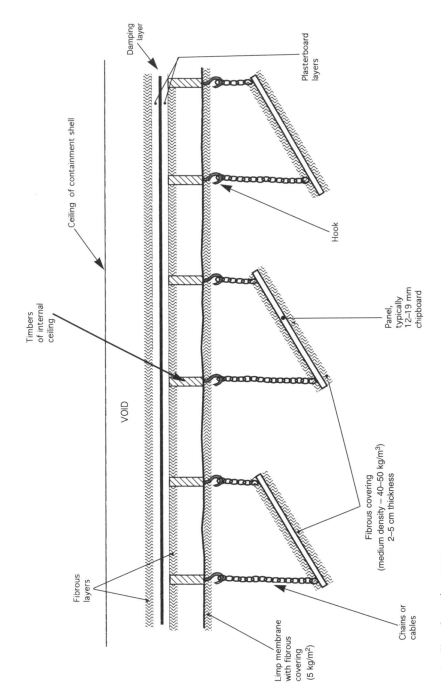

Figure 2.4 Typical ceiling absorption system

Damping layer

Plasterboard layers

Ceiling of containment shell

Hook

Timbers of internal ceiling

Panel, typically 12–19 mm chipboard

VOID

Fibrous covering (medium density – 40–50 kg/m³) 2–5 cm thickness

Fibrous layers

Chains or cables

Limp membrane with fibrous covering (5 kg/m²)

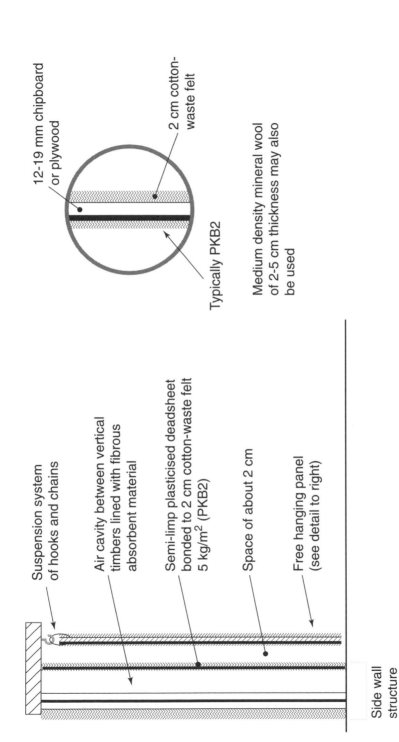

Figure 2.5 Flanking absorbers

12-19 mm chipboard or plywood

2 cm cotton-waste felt

Typically PKB2

Medium density mineral wool of 2-5 cm thickness may also be used

Suspension system of hooks and chains

Air cavity between vertical timbers lined with fibrous absorbent material

Semi-limp plasticised deadsheet bonded to 2 cm cotton-waste felt 5 kg/m^2 (PKB2)

Space of about 2 cm

Free hanging panel (see detail to right)

Side wall structure

however, perhaps we can briefly digress, in order to look more closely at what we shall be seeking to achieve, and why.

2.3.2 Relative merits of neutrality and idiosyncrasy

Neutral rooms are flexible rooms in which work is usually quick and comfortable. An ensemble placed in a neutral room will tend to be heard and recorded with the natural predominances of that ensemble, with the room favouring neither any instrument nor position to any significant degree. However, if this were the be all and end all of recording, this would be a very short book. No, neutral rooms are not the best rooms for all purposes, a point which was perhaps first discovered, at least partially, by accident. There are many studios around which have rooms that were no doubt intended to be neutral, but which have failed to realise that goal. From time to time, a resonance in such a room, or a certain characteristic pattern of reflexions, can produce an enhancement of certain types of music and instruments played in them. They can become great favourites for the performances of those pieces. The same is true for the stages of certain concert halls, and indeed of other halls which have not necessarily been specifically designed for music. Unfortunately, in many of these, a characteristic of the room which enhances the music may only do so in certain keys or at certain tempos, so their suitability for recording becomes more limited.

For example, a symphony played in E major may well be strongly reinforced by a room resonance when certain parts are played with gusto. Perhaps if the coincidence is very fortunate, the main characteristic reflexion patterns will have a natural timing which produce a powerful effect if they coincide closely with a simple fraction of the beats per minute of the tempo. Such a room may give inspiration to the musicians, not only sonically lifting the music, but also encouraging a more enthusiastic performance. These rooms can have their places in both the recording *and* performing worlds in a way which a neutral room may never achieve, but though these rooms may achieve great results in a case such as that stated above, an orchestra performing a symphony in a different key, and with a different tempo, may have difficulties with the above room. If played in the key of F major for example, the resonances around the E may be entirely inappropriate, causing emphasis to notes which should not be emphasised, and masking and weakening the notes which the conductor would prefer to be dominant. In such rooms, for every peak in the response, there will be a dip elsewhere. Furthermore, series of ill-timed reflexions (echoes) may create confusion, and a degree of difficulty with the natural flow of the music. Not all musicians may fully realise what is going on, but many may comment on how they just cannot produce their best in that room with a given piece of music.

This is one of the reasons why much classical music is still recorded outside of studios, either in concert halls (with or without an audience) or in town halls, churches, or similar locations. It gives a choice of ambience for the producer, engineer and conductor to try to achieve the 'ultimate' from selected performances in selected locations. On the other hand, except for some *very* highly specialised recording companies, moving to a different location for each piece of music to be recorded for an album of pieces of music of less than symphonic length would be financially ruinous. What is more, choosing an idiosyncratic studio for the main piece would perhaps seriously compromise the remainder of the album. Such is one very important reason why neutral rooms are so widely used in the parts of the recording industry where high quality recordings must be able to be made on a predictable, rapid and reliable basis. They are especially useful for broadcast studios, where good quality recordings must be made quickly, and at reasonable cost as they are perhaps intended for a once-only transmission.

2.4 Further design aims

The next step in the design of our neutral room is therefore how to add into our relatively dead shell as many desirable features as possible, with as few

Figure 2.6 Decay responses (reverberation)

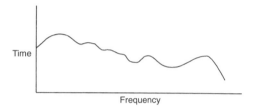

(a) An undesirable irregular decay response. This type of curve will cause colouration of the recordings

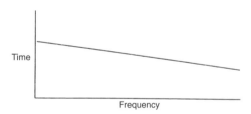

(b) A smooth, desirable decay response showing freedom from resonant colouration

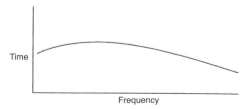

(c) The lower plot would be typical of small rooms with less LF energy

problems as possible. The major pitfalls to be avoided are erratic changes in the reverberation time/frequency characteristic, bunched echoes in terms of their temporal spacing, and strong highly directional echoes (reflexions).

Figure 2.6(a) shows the typical sort of reverberation response that we are trying to avoid, and Fig. 2.6(b) the response which is more in the order of what we are trying to achieve. The response of (a) shows lumps in the curve which are characteristic of unwanted resonances. The peaks are the frequencies which will continue long after an instrument has stopped playing, and the dips represent notes which may appear weak. The irregularities will therefore favour certain notes, suppress others, and mask many low level details of the sound. Such responses at low frequencies are often the result of large, parallel reflective surfaces which can support the strong, axial resonances. It was stated earlier that the angling of the walls away from parallel would help to redistribute the energy in the axial modes, but at low frequencies, the behaviour of sound waves is not always obvious.

In order to reflect at low frequencies, surfaces need to be of a size comparable to a substantial proportion of a wavelength, or the acoustic wave will tend to engulf them and pass around. In our neutral room, we can therefore avoid these problems by placing any necessarily large, reflective surfaces, such as large glass doors or windows, in positions where they do not directly face each other. Other reflective surfaces, necessary for the addition of life to the middle and high frequencies, can be arranged such that they have gaps in between them, the gaps being at intervals of less than half a wavelength of the highest of any troublesome resonances. Alternatively, they can be arranged in suitably random spacings.

2.4.1 What is parallel?

The term 'parallel' in its acoustic sense is very frequency dependent. Figure 2.7 shows two reflective walls, each 10 m long and spaced 10 m apart. They are geometrically parallel, and hence also acoustically parallel at all frequencies. A clap of hands at point X will generate a sound containing very many frequencies, and the sound will propagate in all directions from the source. The waves impinging on points Y and Z will be reflected back through the position of the source, and will continue to 'bounce' backwards and forwards, along the line Y–X–Z. Frequencies whose wavelengths coincide with whole fractions of the distance between Y and Z will go through positive and negative pressure peaks at positions in the room which coincide on each reflexion. They will drive the resonant modes, which strongly reinforce each other, and will tend to be audible in some points in the room but not in others. A 70 Hz standing wave pattern is shown in Fig. 2.8. The light areas show regions of low pressure changes, where the waves would be inaudible, and the dark areas show the regions of high pressure changes, where the 70 Hz content of the sound could be clearly heard.

If we now angle the walls, in a manner shown in Fig. 2.9, with one end of one wall swung in towards the other wall by 1.5 m, we will have two walls with a 15% inclination. Now, a handclap at point X will again send a wave in the direction of Y, which will return to the source point as a reflexion, and will continue on to point Z. A direct wave will also propagate to point Z, and both the direct and reflected waves will reflect from point Z, not back

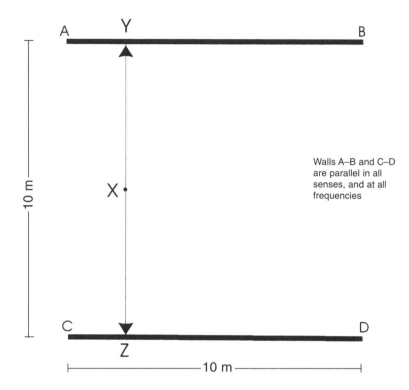

Figure 2.7 Geometrically parallel walls. A sound emanating from point X will spread in all directions, but the sound waves travelling in the directions of points Y and Z will reflect back along the line of their original travel. They will continue to reflect backwards and forwards along the same path, creating echoes, until their energy is finally dissipated by losses in the walls and the air. Such are the paths of axial modes, which, when wavelengths coincide with whole fractions of the distance between the walls, produce modal resonances – see Fig. 2.9

towards point Y, as in the case of the geometrically parallel walls, but towards point F. They will then reflect to point G, and on to point H. Unlike in the case of the geometrically parallel walls in Fig. 2.7, a person standing at point X will not hear the chattering echoes, and most of the resonant energy of the room modes will be deflected into the tangential type, taking a much more complicated course around the room. However, whilst the higher frequencies will be deflected along the pathways Y–Z, Z–F, F–G, G–H, at lower frequencies, where the wavelengths are long, axial modes may still persist. This suggests that at low frequencies, the walls must still be parallel in an acoustical sense.

Figure 2.10 shows the 70 Hz standing wave pattern for a room with one wall angled to the same degree as that shown in Fig. 2.9. The pattern is remarkably similar to the one shown in Fig. 2.8. Although Fig. 2.9 shows that the angling of the walls has produced a very different path for the hand-clap echoes, and will be quite dispersive at high frequencies, at low frequencies very little has changed. Essentially, for geometric angling to be acoustically

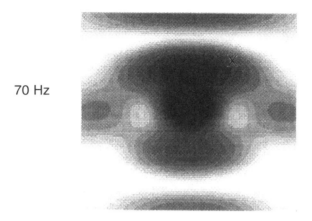

70 Hz

Figure 2.8 Magnitude of pressure field due to a point source between the two walls depicted in Fig. 2.7

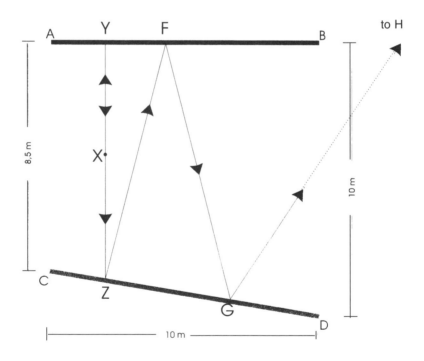

Figure 2.9 Here, the general situation is that of Fig. 2.7, except that one of the reflective surfaces has been moved to create a geometrically non-parallel arrangement between the surfaces A–B and C–D. Distinct echoes, as produced by the surfaces in Fig. 2.7, are not possible, as the reflexions will not follow repeating paths. Mid- and high-frequency sounds emanating in the Y and Z directions, from position X, will subsequently follow the path Z, F, G, H etc. At low frequencies, however, all may not be too different from the conditions of Fig. 2.7 – see Fig. 2.10

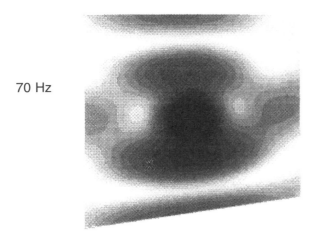

70 Hz

Figure 2.10 Magnitude of pressure field due to a point source between the walls depicted in Fig. 2.9

effective, the pathlength differences for subsequent reflexions must be a significant part of the wavelength. With the wavelength of 50 Hz being about 8 m, the degree of angling required to be acoustically non-parallel would perhaps be possible in buildings the size of concert halls, but would be likely to consume too much potentially usable space in a conventional recording studio.

The effect of a limited degree of wall angling on the response of a room is shown in Fig. 2.11. The two traces show the performance of the walls shown in Figs 2.7 and 2.9. Above 300 Hz, the non-parallel walls show a clear reduction in modal energy when compared with the parallel walls, but below about 100 Hz there is little difference between the two traces, showing that, in the

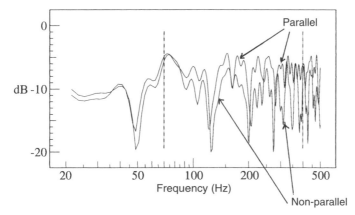

Figure 2.11 The above plots show the response at point X in Figures 2.7 and 2.9. It can be that, at frequencies below around 80 Hz, the effect of the angling of one of the reflective surfaces has had only minimal effect. Above about 200 Hz however, the effect is quite pronounced.

acoustic sense at least, the walls in Fig. 2.9 are still parallel. The reduction in modal energy above 300 Hz is largely due to the fact that the wall angling drives more of the higher frequency modes from the axial to the tangential type. The tangential modes not only have more complicated paths to travel, but also strike the walls at oblique angles, which tends to rob them of more power than is lost in the more perpendicular impacts of the axial modes. It can thus be seen that whilst the angling of walls can have a very worthwhile effect at frequencies above those which possess a wavelength which will be subsequently shifted in position by a half wavelength or more on their return to the source wall, at frequencies *below* these, the effect will simply be that of a comb filter, as shown in Fig. 2.12. Here, as the frequencies are swept downwards from the above half wavelength frequency, the sweep passes through alternately constructive, neutral and destructive regions. Strong comb filtering at low frequencies is usually musically disruptive and very undesirable in recording studios and music rooms in general, though it exists to some degree in all reflective spaces. At higher frequencies, our ears use it for many beneficial purposes, such as localisation and timbral enrichment.

So, whilst the angling of reflective surfaces is a viable technique for mode control at middle and higher frequencies, at low frequencies, geometric solutions are usually not able to produce the desired results, and hence absorp-

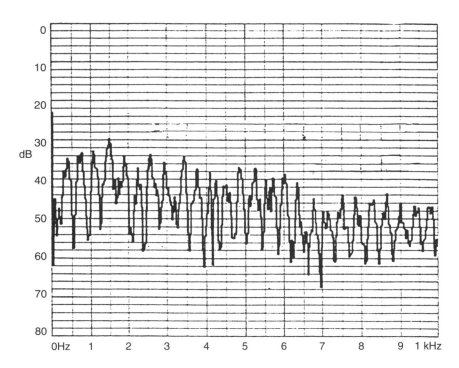

Figure 2.12 The averaged power spectrum of a signal with one discrete reflexion. Comb filtering is revealed clearly on a linear (as opposed to a log) frequency scale, where the regular nature of the reflexion-produced disturbances can clearly be seen. In the instance shown above, the additional pathlength of the reflected signal over the direct signal was just under 10 m, producing comb filtering with dips at a constant frequency spacing of just under 40 Hz.

tion must be resorted to, though diffusive techniques are beginning to become a practical reality. These will be discussed later in the chapter. Parallel surfaces also produce the repetitive chatter or 'slap back' from impact noises, somewhat akin to the ever-receding images seen when standing between parallel mirrors, and this reflective chatter can be equally as undesirable as the resonant modes in the destructive effect on the music. In the following sections, we shall begin by looking at practical solutions to the circumvention of these problems, whilst producing a desirably neutral acoustic.

2.4.2 Reflexions, reverberation and diffusion

We are now faced with the problem of how to put the ideas so far discussed into a practical form. Unfortunately, there are so many ways of doing this that a whole book could be filled on this topic alone, so we will have to take an approach which will incorporate a number of solutions in one design. From this, hopefully, it will be possible to gain something of a feel of the range of possibilities open to designers, and the way that these can be put into general recording practice. What is necessary for musical neutrality in rooms of the size under consideration here (650 m^3) is a reverberation time (or, in these cases, more correctly a decay time) in the order of 0.3 to 0.5 seconds, perhaps rising to 0.7 or 0.8 seconds at low frequencies, with the rise beginning gradually below 250 Hz. Figure 2.13 shows a typically desirable frequency/time decay response of such a room; the time indicated being that taken for a sound to decay to 60 dB below that of its initial level.

There are two general techniques involved in creating reverberation in such a room; either by reflexions or by diffusion. In recent years, companies such as RPG[1] in the USA have created a range of acoustic diffusers capable of operating over a wide range of frequencies. These are constructed on principles based on sequences of cavities whose depths alternate according to strict mathematical sequences. They can be made from any rigid material, but perhaps wood, concrete and plastics are the ones most frequently encountered. The mathematics were proposed by Professor Manfred Schroeder[2,3,4,5] in the 1970s, and are based on quadratic residue sequences. The effect of the cavities is to cause energy to be reflected in a highly random manner, with no distinct reflexions being noticeable. The random energy scatter creates a reverberation of exceptional smoothness. By using such diffusive means, the

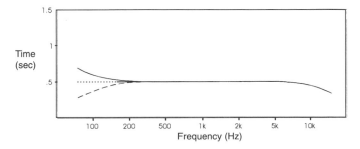

Figure 2.13 RT$_{60}$ of a good neutral room. Low frequency reverberation times can be allowed to vary with room size

overall reverberation time can be adjusted by the ratio of diffusive surfaces to absorbent surfaces, though for an even distribution of reverberation in the room, a relatively even distribution of diffusive surfaces is required. Except for the floor and windows, all other surfaces are usually available for diffusion.

With the availability of diffusers, achieving the desirable degree of ambient neutrality may well at first seem to be a simple matter of adding diffusers until the desired reverberation time is achieved, but from such a room there would usually be a lack of musicality. What such a room would lack are discrete reflexions. Fortunately, they are easily introduced, as they are of great importance to both the musicians, and their audiences. On concert stages, they are needed by the musicians to reinforce the sound of any instrument or ensemble. As mentioned earlier, reflexions divide into two groups, late reflexions and early reflexions. The early reflexions, arriving less than 30 or 40 ms after the direct sound, are heard by the ear as a timbral enrichment of the instruments. The later reflexions, arriving 40 ms or more after the direct sound, add spaciousness to the sound, which for many types of music is essential for its enjoyment. The dimensions of our example of a large neutral room were chosen such that an instrument in the centre of the room would produce only late reflexions from the wall surfaces, but of course, from the floor and ceiling, the reflexions would necessarily be of the early type. Instrument to floor distances are normally relatively constant; floor reflexions typically being in the 5 to 10 ms region. Any ceiling reflexions in this room would be in a borderline 20 to 40 ms region, dependent not only upon ceiling geometry, but also upon whether diffusive, reflective or absorbent surfaces predominated. In later chapters dealing with more reverberant spaces, we shall look at reflexions further, but in the neutral type of room under discussion here, whatever reflexions exist should tend to be reasonably well scattered, otherwise they will develop a character of their own, and the room would lose its neutrality.

2.4.3 Floor and ceiling considerations

Let us now look at possible ceiling structures for our neutral room. As we discussed earlier, the nature of the floor has been chosen to be wood. Carpet tends to produce a lifeless acoustic, uninspiring for the musicians, and unhelpful for the recordings. Stone was rejected, partly for its 'harder', more strident reflective tendency, but also on the practical ground of slippage. Floors of neutral rooms overwhelmingly tend to be of wood. At this point of the design stage (see Fig. 2.2(b)) we have a very dead ceiling at a height of about 4 m 50 cm. We cannot come down too much below this, as the reflexions which we would introduce would tend to become of the tone-colouring, early nature. What is more, too low a ceiling would preclude the siting of microphones above any instruments at a height which may be necessary to cover any given section of musicians.

One solution to such a problem is to construct a ceiling of wooden strips, with spaces in between them, which will allow a good proportion of the lower frequencies to pass into the absorbers behind. They therefore provide mid and high frequency reflexions, but without allowing an unwanted low frequency build-up. As mentioned earlier, in order to reflect at low frequen-

Figure 2.14 Various ceiling constructions for a neutral room

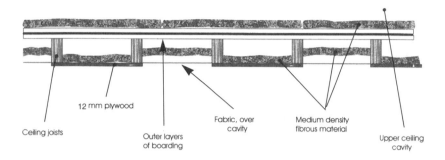

12 mm plywood

Ceiling joists

Outer layers
of boarding

Fabric, over
cavity

Medium density
fibrous material

Upper ceiling
cavity

(a) Alternate hard and soft surfaces

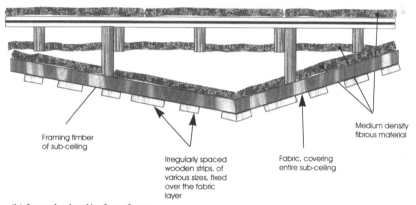

Framing timber
of sub-ceiling

Irregularly spaced
wooden strips, of
various sizes, fixed
over the fabric
layer

Fabric, covering
entire sub-ceiling

Medium density
fibrous material

(b) Irregular hard/soft surfaces

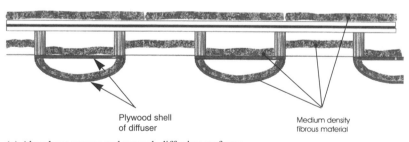

Plywood shell
of diffuser

Medium density
fibrous material

(c) Absorbent spaces and curved, diffusive surfaces

cies, surfaces need to be of sizes comparable to the wavelengths to be reflected. We can therefore have some degree of control over the lower limit of our desired reflectivity by providing gaps in our reflective surfaces at appropriately chosen intervals. Once again, with this complex subject, there are so many ways of achieving each objective, but here I shall describe some of the possibilities within one technique of specific interest. The strips to be described could be of hardwood or softwood, and could be plain, varnished, painted, rough or smooth. Each will give its own subtle character to the sound. Hardwoods, untreated, can produce quite interesting sounds, but in today's ecological climate, I personally refuse to specify any woods of an exotic or not too easily replenishable nature. Sound is of course important, but we do have a planet to protect as well.

Obviously, we do not want to encourage the build-up of resonant energy in any modes which would unpleasantly colour the sound, so the ceiling surface can be broken into a series of angled sections, set to produce what the designer would consider to be the most appropriate angle of reflexions. A selection of possibilities for ceiling designs is shown in Fig. 2.14. Figure 2.15 shows a pattern of wood/space ratios which is based on a numerical sequence, not dissimilar to the one used for the diffuser cavities referred to earlier, and indeed there *is* a certain amount of diffusion from this type of arrangement, which is partly the result of diffraction from the edges of the wooden strips. The purpose of this arrangement, however, is to help to prevent any noticeable patterns forming in the reflected sound-field.

The BBC[6] (British Broadcasting Corporation) have developed a very successful range of ceiling tiles which fit into the typical sort of false-ceiling structures used in many offices and broadcast studios. In fact, using a steel-framed grid for a false ceiling allows great flexibility in the introduction of absorbent, reflective or diffusive tiles, which can be an excellent way of providing a room with a degree of acoustic variability. In the broadcasting world, these systems are widely used, but for my own tastes, in commercial music recording studios, their appearance is too industrial to provide what I consider to be the necessary decorative ambience for musicians to feel comfortable and creative. However, I realise that these things are highly personal, and for people who find these prefabricated tiles pleasing to look at, they can be safe in the knowledge of their excellent acoustical performance.

2.4.4 Wall treatments

Figure 2.16 shows a possible wall layout for our neutral room, and it can be seen from the figure that parallel wall surfaces have been avoided. Where the frequencies begin to render the walls acoustically parallel, they are allowed to pass into absorbers, either directly or after first reflexion. Care has been taken to make sure that glass windows or doors do not face directly towards any parallel reflective surfaces. In the room shown, the control-room door/window system is set into a relatively absorbent wall, so one of the walls is now defined.

That now leaves us with three wall surfaces to complete. For musical neutrality, we do not want too much reflective or reverberant energy, but just enough to give the room sufficient life to stop the instruments from sounding too dead. It is similar to a little seasoning, bringing out the flavour of the food without overpowering it. The glass surfaces, the entire floor, and the ceiling reflexions are more or less enough for this. We also have the problem that if we make the walls reflective to any significant degree, we may create unduly reflective areas for any musicians playing close to the walls. In fact, in the corners of the room the colouration could become most unnatural if bounded by the floor and two reasonably reflective walls, all

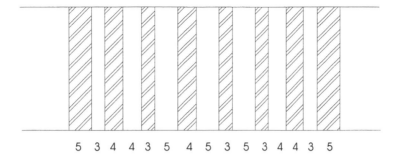

5 3 4 4 3 5 4 5 3 5 3 4 4 3 5

Figure 2.15 Typical arrangement of wood strips and absorbent openings. The widths of the strips and openings are based on a numerical sequence that provides more or less equal areas of absorbent and reflective surfaces, but without any simple regularity which could lead to problems at certain frequencies that may have properties coinciding with the regularity. The numbers in the diagram represent the relative dimensions of the adjacent strips and openings

providing early reflexions in addition to those from a borderline early/late ceiling.

For reasons of structural integrity if we have walls covered with fabric for decoration, we need wooden rails, at waist height or a little higher, to prevent people from 'falling' through the fabric. We need skirting boards so that the floors can be cleaned without soiling the fabric, and we may also need to put rails at shin height as a convenient mounting for microphone sockets and electrical outlets. Figure 2.17 shows a small version of a room with such fittings, but should such a room be considered to be just a little too dead, or in cases where only a rather small area of glass is to be used, then an arrangement such as that shown in Fig. 2.18 can be employed. The reflective surfaces are based largely on the concepts of Fig. 2.15, but the ratio of the areas of the spaces to the facing timber can be adjusted to produce the required degree of acoustic 'life'. The fabric covering is for decorative purposes only, and so should be of a type which is acoustically transparent. Many fabrics can be surprisingly reflective, and if stretched too tight for neatness, they may well also act like drum skins. Lightweight 'stretch' fabrics are useful here.

So now we have something approximating to a musically neutral room, which tends neither to add character to the sound of an instrument nor 'suck out' all of its life. Rooms such as the ones being described here are excellent for recording efficiently, rapidly and predictably, but if they are the *only* spaces available for recording, then the practice still harks back to the philosophies of yesteryear, and their more 'technical' correctness. Perhaps these rooms had more of a place when control rooms were less precise, when monitoring conditions were less predictable, and when the electronics of the recording chain itself were more coloured in their sound character. Perhaps it is now a good time to develop these thoughts further, and look at the alternatives. Remember, though, neutral rooms are very necessary in the broadcast industries, and in recording facilities where rapid results are necessary from a wide range of music and instruments. For these purposes, they excel.

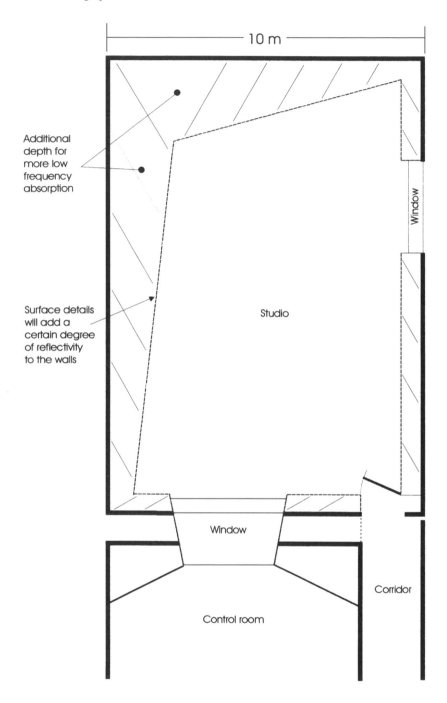

Figure 2.16 Possible layout of neutral room

Figure 2.17 A relatively small room with a very neutral acoustic characteristic. Studio room of the Ukrainian Air Force, Cultural, Educational and Recreational Centre, Vinnitsya, Ukraine

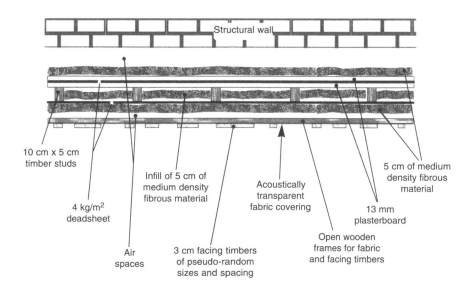

Figure 2.18 View from above, of a wall structure and treatment for a totally 'neutral' room

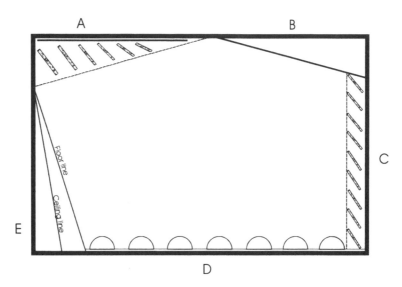

Figure 2.19 Possible layout of a neutral room with a little flexibility (doors and windows omitted for simplicity). The ceiling could be typically one of those shown in Fig. 2.14.
The walls:
A Low frequency absorber with controlled upper mid/higher frequency reflectors. Similar surface features to those shown in Fig. 2.17
B Wood-panelled, reflective wall
C Wide-band absorber, faced with fabric-covered frames
D Diffusive wall, with absorbers between diffusing half-cylinders
E Reflective, double-sloped wall, faced with wooden panelling

As further food for thought, Fig. 2.19 shows a neutral room with a some-what more ambitious approach. It is neutral on the 'macro' scale, but on a more 'micro' scale, in close proximity to the different wall surfaces, a degree of variability of acoustic performance can be achieved. This room therefore straddles the borders of neutrality and variability, so perhaps that will lead us nicely into the next chapter.

References

1 D'Antonio, Peter, 'Two Decades of Diffusor Design and Development', AES Pre-print, 99th Convention, New York (1995)
2 Schroeder, M. R., 'Diffusive Sound Reflection by Maximum Length Sequences', *Journal of the Acoustical Society of America*, Vol 57, No 1 pp 149–150 (1963)
3 Schroeder, M. R., 'Comparative Study of European Concert Halls: Correlation of Subjective Performance with Geometric and Acoustic Parameters', *Journal of the Acoustical Society of America*, Vol 56, No 4 pp 1195–1201 (October 1974)
4 Schroeder, M. R., 'Progress in Architectural Acoustics and Artificial Reverberation in Concert Hall Acoustics and Number Theory', *Journal of the Audio Engineering Society*, Vol 32, pp 194–203 (1984)
5 Schroeder, M. R. and Gerlack, R., Response to 'Comments on Diffuse Sound Reflection by Maximum Length Sequences', *Journal of the Acoustical Society of America*, Vol 60, No 4 p 954 (October 1976)
6 Walker, R., 'The Design and Application of Modular, Acoustic Diffusing Elements', Proc of Institute of Acoustics, Vol 12, Part 8 pp 209–218 (1990)

Chapter 3

Variable rooms

Chapter 2 was rather long and convoluted, but it had to be in order to set out the groundwork for the expansion of our design concepts. I described in considerable detail rooms which, due to their limitations, cannot be the 'all things to all people' that they were perhaps originally intended to be. Such rooms are excellent for recording jingles, one-off radio broadcasts, and many other recordings which must be cost-effective and of good quality. They are real 'work horses'. Do not think that I am relegating these rooms to a second division, as I am not. They are, however, not the ideal rooms for many of the 'big production' recordings for CD release, which will be listened to countless times, frequently by people with expensive equipment and critical ears, and from which 'special' sounds have come to be expected. One of the main failures of large neutral rooms is not that they do not produce good recordings, but that they frequently do not inspire the musicians or help in the production of 'magic' sounds. As mentioned in Chapter 1, they were originally conceived at a time when it was the 'accepted' view that the engineers recorded just what the musicians played. The studio rooms, themselves, had not yet been fully integrated into the creative process of music recording.

In the late 1960s, such neutral rooms as those described would have been well received by the majority of recording personnel, but around the same time, groups like The Who, Led Zeppelin and the Rolling Stones were beginning to use the old Olympic Studio in Barnes, London, which had a large recording room of distinctly characteristic acoustics. Perhaps for the first time, artistes and producers were beginning to get more general freedom relating to how, where and when they recorded. They were becoming less tied by the instructions of their record companies. The 'supergroups' began to emerge, having so much money that they could even start their own record companies, and, not *too* much later, have their own front-line studios built. It is true that in the late 1950s and early 1960s, visionary producers like Phil Spector in the USA and Joe Meek in the UK were innovative and controversial, but these people were very rare exceptions, and faced much opposition from the 'establishment'.

3.1 Time for change – musician power

In times gone by, the technical people often seemed to look down on the musicians, considering them pawns in the 'scientific' recording process. Given their new found independence, the musicians began to make their point 'with their feet', walking out of the studios where they felt uncomfortable, and into the studios where they felt they could play their music to the best of their potential. Time after time, the newly 'in-vogue' studios were those which had been frowned upon by the more mainstream acousticians and recording staff. Yet, ironically, it was very often the technical *weaknesses* of these studios which the musicians found that they could use to their advantage.

Sounds in general were becoming a more important part of the recording process. 'Records' were becoming more of a complete creative process. Rather than a studio just recording what the musicians played, they were becoming more of an instrument in themselves. This was somewhat different to what had begun in the late 1950s, when at that stage it was mainly the electronic processes that were beginning to be usurped by the adventurous producers and musicians. The studios were still largely sacrosanct. Studio staff in the more 'comfortable' studios were also often of a more flexible nature, with maintenance staff who were a little artistically inclined themselves, or at least who were interested in new ideas and experimenting with novel techniques. It was a long time however before the old attitudes were swept away, partly because there was a certain 'superiority' seen in being 'technically correct'. I am viewing this at the moment from a European perspective, as in the USA, greater emphasis always seems to have existed on the subject of musical acoustics. In the 1960s, many British groups began to go to the USA to get the sounds that they sought, with few, if any, American groups coming to Europe specifically to record, and when they did, it was usually for the European musicians and not the studios.

It was clearly becoming apparent that acoustic conditions, other than musically neutral ones, were being more widely appreciated, but all too frequently the conditions were idiosyncratic. They were great for a certain type of recording, but imposed their character too much on some recordings which did not need that 'help'. It was thus also becoming clear that a greater degree of acoustic variability was needed. What was more, that variability needed to be something more than the provision of acoustic screens, which at the time were usually on wheels, and *perhaps*, if one was really lucky, had one side hard and the other side soft. It was also becoming no longer acceptable to enclose the drummer in a box, crudely arranged from screens, solely for the purpose of achieving better separation from the other instruments (and vice versa).

Unfortunately, just as this necessity for acoustic variability was beginning to be more widely appreciated, 16-track recording was beginning to become widespread, followed in 1972 by the first practical 24-track machines from Ampex and MCI. (A Unitrack 24-track machine went into Morgan Studios in London in 1970, but reportedly was never fully operational.) The new vogue word was 'separation', and for the first time all the instruments of many groups, and even the individual drums in a drum kit, became recorded on separate tracks. This supposedly offered much more scope for processing in the mixing stage, but inter-instrument leakage from track to track could limit some of the possibilities. In prior times, musical errors would usually require a re-take of the whole backing track, but with multi-track recording, single instrument performances could be 'repaired'. *However*, if, for example, a guitar needed to be replaced, then even if the old track was erased, if the original had spilled over into the drum microphones, the unwanted guitar, which may have played out of time in the original, or played wrong notes, would still be audible via the drum mix. The repair of the original guitar would thus only be undetectable if the leakage between microphones was minimal, and that meant either recording in dead spaces or in multiple, isolated rooms.

The next general phase in variability, not surprisingly, was to provide a main recording area, with one or a series of 'isolation rooms', usually in good visual contact with the main area. Vocal booths had existed for a long time, but these new isolation rooms were usually larger. In response to the need for more variable acoustics, they were often built with different characters, having varying degrees of life (both acoustically, and before (in many cases) rejection by the musicians).

Musicians rarely like being isolated: it frequently makes it difficult for them to play as one unit. Perhaps the most emphatic of all of the musicians who rejected these ideas were the drummers. Technological advances and the perceived benefits of greater isolation led to the introduction of the disastrous drum booths of the mid-1970s. Engineers and producers saw much possibility in the prospect of more isolated drums and a dead environment, but drummers usually hated playing in such dead surroundings. This will be discussed in more detail in Chapter 5, but suffice it to say here that there was a period of several years of experimenting with multi-track recording and high degrees of separation. Once again, largely due to a revolt by the musicians, the pressure was back on to produce rooms with a good ambience for playing (both visually and acoustically) and where the musicians felt happy, yet where the recording staff could also achieve their aims. Larger rooms began to arrive on the scene with either zones of different acoustics, or movable walls and/or ceiling surfaces which could change the ratio of absorbent, diffusive or reflective surfaces facing the room. From time to time, movable

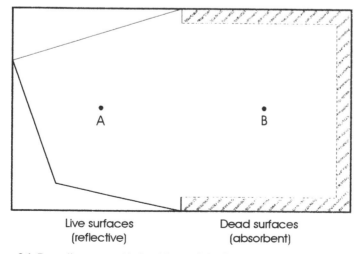

Live surfaces   Dead surfaces
(reflective)   (absorbent)

Figure 3.1 Recording room with fixed live and dead areas

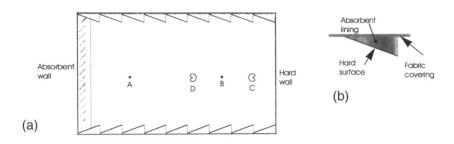

Figure 3.2 (a) Recording room with graded acoustics. (b) Detail of side wall

ceiling panels could be raised and lowered, thus being able to change their reflexion patterns from the late to the early type.

Figures 3.1 to 3.3 show three possible ways of achieving acoustic variability in larger spaces. Although the type shown in Fig. 3.1 is relatively cheap to construct, and highly effective, it has the drawback of being awkward to use if a large ensemble is in the room, as a uniform acoustic could not be shared by the whole group. Figure 3.4 shows how the room could be used effectively for a typical rock group, allowing the musicians to play as one unit in close contact with each other, yet with each instrument in its own, desired, acoustic space. The results can be excellent, but rarely are large rooms dedicated solely to the recording of a small number of musicians. Unfortunately, if a larger ensemble is placed in such a room, filling more than one half of the space, either the front/back or left/right balance of direct to reverberant sound will not be uniform. It would simply not be possible to achieve a balanced, overall sound. However, in small rooms where the recording of acoustic ensembles was never likely to occur, this technique can be used to good effect. What is more, if a room is large enough, two distinct areas of differing acoustic nature can be very useful. In the area between

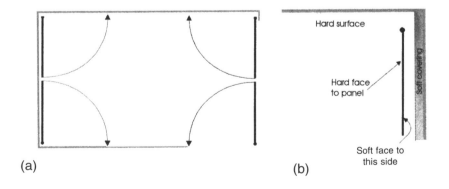

Figure 3.3 (a) Recording room with adjustable acoustic panels. (b) Detail of hinged panel system

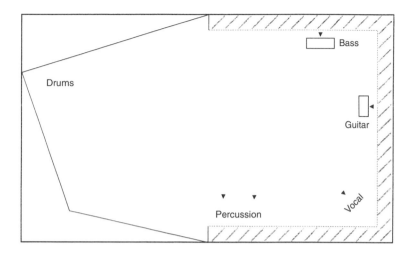

Figure 3.4 Typical usage of room shown in Fig. 3.1 by a rock group. In this example, a five-piece group could record 'live' using the characteristics of the room to good effect. The drums could be set up in the live area, to produce a full sound and a good feel for the drummer. Bass and guitar amplifiers face the absorbent walls, thus reducing the overspill back to the other instruments. The amplifiers help to shield the microphones from the drums and percussion. The percussionist faces the other musicians, and the percussion microphones do not point in a direction where they are likely to collect excessive overspill, even though the percussion is in a relatively live part of the room. The vocalist, in a dead corner, faces the other musicians, but the directional vocal microphone faces the absorbent surface

the two halves, an interesting area can exist where very different recordings can be achieved by the varied positioning of directional microphones, either in the vicinity of the instrument, or more distantly as ambient microphones. In the live end of the studio, walls can be made quite reflective, with angles chosen which will encourage lateral reflexions, and resonances of lowish Q. This area would contain far more reflective surfaces than the neutral room of the previous chapter. The live end of the variable room under discussion here would not be musically neutral, but it would not be intended to be so.

We must always give due consideration to how musicians feel in the recording spaces, because if they are not comfortable, they tend not to play so well, and if they are not playing well then it is barely worth recording their performances. Much music is based around instrumentation, and instrumentation is often based around the sounds that they produce in certain spaces. The evolution of music, instruments, performing spaces and human perceptions has been inseparable. Instruments often rely on the fact that they will be surrounded on at least three sides by walls. Side walls produce lateral reflexions, and those reflexions in effect become part of the sound of the instruments. For this reason we cannot, as was found in the high-separation era, take musicians into alien environments, expect them to perform optimally, then hope to add whatever is necessary later. *An inspired performance must occur at the instant of the recording.*

Certainly for most acoustic instruments, lateral reflexions are expected by the musicians; they are critical to the development of the expected tonal character of the instrument. Furthermore, those reflexions must be lateral; they cannot be adequately substituted by reflexions from other directions. This is one reason why acoustically neutral rooms so often fail to inspire, and why they failed to gain wide appreciation for serious music recording. In a more lively room, we need to provide some reflective wall surfaces to add richness to many instruments, but, on the other hand, for recording of electric bass guitar and amplifier, such surfaces can steal so much of the punch from the sound. What we need to do, to add greater flexibility, is either make the reflective surfaces in some way removable, or confine the bass to a non-reflective area. Lateral reflexions are also a problem as far as separation is concerned, as if we need to collect them in the recording microphones, we may also collect the reflexions from other instruments in the same room. This begins to highlight the need for variability, as what is excellent in one set of circumstances can be detrimental in another.

3.2 Introducing practical variability

Figures 3.1, 3.2 and 3.3 show some steps along the road to variability. In the first case, Fig. 3.1, position A would enjoy a rather live acoustic, whilst position B would be much more dead. A is surrounded on three sides by reflective walls. B is surrounded on three sides by very absorbent walls, and the only live surface directly facing it is the far wall of the live area, which is quite some distance away. Figure 3.2 again has positions A and B, live and dead respectively, though this time the method of achieving the effect is rather different. The sawtooth arrangement of the reflector/absorber surfaces creates a situation whereby as one moves from the absorbent wall to the reflective one, there is a progressive reduction in the reflective surfaces facing the signal source. If an instrument is at position B, with a cardioid microphone at position C, facing B, then an almost dead room response would be recorded, as all the reflexions would be passing away from the front of the microphone diaphragm. Conversely, by positioning the microphone at D, many reflexions would be captured by the microphone, both from the hard wall and from at least the first three 'sawtooth' reflectors of each of the side walls.

Figures 3.1 and 3.2 are acoustically variable in terms of both the positions of the sources of sound and the microphones, but Fig. 3.3 shows a move toward a more variable room. In this instance, hinged panels can be moved through 90°, exposing either hard or soft surfaces to the room. By this means, either the whole room, or different sections of it, can be radically altered between live and dead. I have built many small rooms of this type, along with a variant where the hinged panels are in the centre of the long walls (see Chapter 8: Fig. 8.12 and Section 8.6), allowing the rooms to be sub-divided. They have generally been very well received, but in the case of larger rooms, this type of system is somewhat impracticable.

Figure 3.5 shows a proposal for a large room with quite comprehensively variable acoustics. From this it can be seen that a large room with seriously variable acoustics does not come cheap. Space is consumed by the variable elements, which means that considerably more floor area is needed in the building shell than will be realised in the final recording room. Highly variable rooms tend to be both structurally complicated and expensive, but they

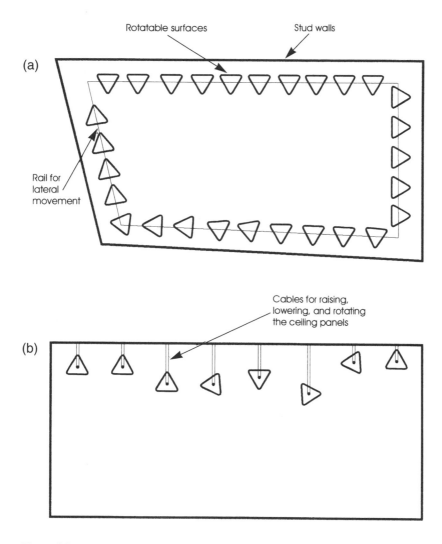

Figure 3.5 Room with high degree of acoustic variability (but at a cost of much floor area). (a) Plan; (b) side elevation

can also be very effective recording tools when a truly multi-functional room is needed. Essentially, though, it will be seen from Fig. 3.5 that almost every surface of the room needs to be capable of being changed from hard to soft if a very high degree of variability is to be achieved. An added possibility would be the provision of carpets, which could be laid down over the floor, or removed, as necessary.

Figure 3.6(a) shows the details of the rotating panels of Fig. 3.5, with four different designs, though many other variations on this theme are possible. The diffusive sides show the options of either curved surfaces of different radii, or the quadratic residue types of both the pit and relief forms. This type of room variability technique is now widely used in concert halls, though it

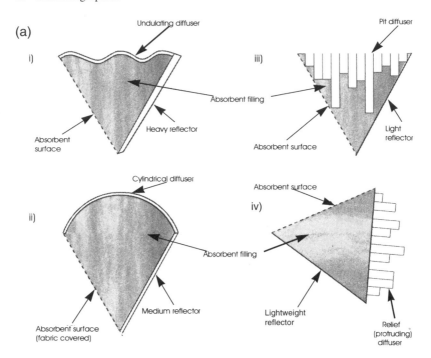

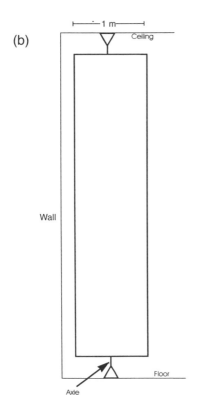

Figure 3.6 (a) Detail (end views) of rotating panels – four variants. Rotating contrivances such as these can provide reflective, diffusive or absorbent surfaces, either wholly, or in combinations, by being rotated to intermediate positions. They can be applied to walls or ceilings, and can be motor driven and controlled from the control room, where their effect on the recording acoustics can be judged whilst listening via the microphones. (b) Mounting arrangement for rotating panels shown in (a). Ceiling mounted devices would use a similar arrangement, but would be mounted horizontally

was largely the Japanese designer, Sam Toyashima, who brought it to high profile in recording studios around the world. The wall panels of Fig. 3.6(a) have three surfaces, which can be rotated into any position desired. Having three surfaces gives the option of mixing any of the diffusive, reflective or absorbent characteristics by intermediate positioning. The room can be divided into live or dead areas, and, by the rotation of the panels, the reflexions are to a large degree steerable, which can create interesting effects for ambience microphones. The ceiling panels can be raised, lowered or rotated, and therefore can be absorbent, diffusive or reflective with varying degrees of timing for first reflexions.

With rooms such as this, it is often surprising to the uninitiated just how much area needs to be changed to achieve any significantly noticeable change in the general acoustics. Except for microphones and instruments in close proximity to the variable panels, rotating only three or four of them would usually be virtually unnoticeable in the context of a change in the overall acoustic of the room. It really takes a change of 20–25% of the overall surface area of the room to have any readily noticeable general effect. In a 15 m × 10 m × 5 m room, accepting for now a hard floor, it means changing something around 100 m^2 of surface area for a worthwhile change in acoustics. Once over that threshold, however, the effect, in which ever direction it is acting on reverberation time, begins to develop rapidly.

The rotating wall-section technique takes up more permanent space than the method of varying the wall surfaces by movable panels, either hinged, as in Fig. 3.3, or attached by some sort of hook system, but with the panel system it is more difficult to achieve the intermediate situation, and it may also be less rapidly adjustable. Obviously, the hinged panels need a free space to swing out, which is perhaps not too much of a problem in a very large room, but in smaller rooms it could mean dismantling a whole drum kit just to make an experimental change. On the other hand, if the changes of acoustics can be pre-planned, the hinged/hung panel system does cause less *permanent* loss of space inside the isolation shell than does the rotating system of Fig. 3.5. As usual, compromises exist in each case.

3.2.1 Small room considerations

Small recording rooms, of less than around 150 m^3, tend to be more difficult in terms of variability. They tend to flip-flop from one state to another, passing through some strange sound characteristics in the intermediate stages. What is more, in a small room the number and positions of people and equipment occupy a much greater proportion of the room volume, and hence may themselves have a great bearing on the acoustics. The variations in the surfaces of small rooms usually have to be judged once the room is ready to record, with all personnel and equipment in place. As mentioned earlier, with wheeled or hinged panels, changing things after the instruments have been positioned can become a very disruptive and slow process in a congested room. I usually prefer to sub-divide smaller areas into sections whose acoustics are likely to be the ones most appropriate for the majority of the music which the studio expects to be recording. A technique such as that shown in Fig. 3.5 would be unlikely to be a good choice. The overall loss of space due to the movable sections would be a much greater proportion of the space in the isolation shell, and their effects would not be so subtle because they would be very close to the instruments or vocalists.

The important point to remember is that the movable panels of Figs 3.5 and 3.6 will not scale. They cannot be built one half of the depth in a room of one half of the volume. The effects of the panels are related not to room sizes but to wavelengths, and as the frequency ranges of the instruments in a small or large room remain the same, some of the dimensions of the variable wall sections must remain the same. If an absorber needs to be 1 m deep, then in a room of half the total surface area, although only half the surface area of absorbers may be needed, they still need to be 1 m deep. I have been in studios where small scale versions of these devices have been tried, but they have tended to be effective over much narrower frequency bands, and their sonic effect has usually been very unnatural. For so many reasons, recording spaces tend to be like boxers; a good big one will almost always beat a good little one.

In the smaller rooms, it is almost impossible to achieve situations whereby the first reflexions back to the instrument are of the late type (echoes). There are many more early reflexions returning to the musicians and the recording microphones, so the reflexions of this type produce tonal colouration rather than a sense of spaciousness. This can be off-putting for the musicians, whether they hear the sound directly or via the microphones and fold-back systems, both of which can affect the way that they play. All in all, it has been my experience that smaller rooms are better suited either to fixed design concepts, or to being provided with the means of gross state changes. Subtle variability rarely seems to achieve its aims in small spaces, although proprietary diffusing systems, such as those produced by RPG, do offer solutions to many of the small room problems. They can at least provide a degree of reverberant life, without a build-up of troublesome early reflexions, but in situations which *need* reflexions for spaciousness, they can do little to help. Even diffusers suffer from wavelength limitations, and for good low frequency diffusion, they currently need to be relatively large in their front-to-back dimension. However, some interesting research is afoot, whereby actively driven end-walls, at the back of the pits, may be able to simulate in a shallow pit the effect of a deep one. Time will tell! Nonetheless, in experienced hands, the idiosyncratic nature of small, variable rooms can be put to good use, especially in the more 'creative' environment of modern music. In a situation where 'different' sounds are being sought, there can be some interesting possibilities, but one has to be careful that similar room-sound characteristics do not build up on themselves to become overbearing in the final mix. The following story may help to highlight the point.

I was once involved in the mixdown of a tape containing five acoustic guitars from a television recording. The guitars had been recorded via pick-ups on a very large, important live recording before an audience of 10,000 people. During the mixdown for TV, it was realised that the spirit of the occasion had somewhat dominated the event. Of the five guitars, only two were generally usable, but nonetheless, the programme looked great. The band in question were platinum sellers in their home country, and it was not long before their record company began pressing for a live album from the event. Well, what will stand for a one-off TV broadcast and what will stand critical repeated listening, without the distraction of a picture to watch, are two different things. It was decided that all five guitars would need to be replaced. This is not quite the musical travesty that it appears, as it was something of a celebration concert, with rather an excess of guitars, but the record company wanted to keep the overall feel of the live event which had an unusually 'electric' atmosphere. The rhythm track was excellent and the vocals were powerful, so the job seemed worthwhile.

Using microphones for recording the acoustic guitars at such a live event was not practical, but on the studio albums of the band, the guitars had all been acoustically recorded. A studio was booked to make the overdubs, and it was decided to record the replacement guitars with microphones, and not with the pick-ups used in the live recording. The engineers recording the replacement guitars chose a position in a relatively live room where they considered the correct balance to exist between the direct and ambient sound, giving a fullness to the guitars which the direct injected recordings lacked. Although each guitarist brought his own guitar, when it came to the mixing stage it became painfully apparent that all the recordings had been done in the same place in the same room; and what was more, with the same microphones. Each guitar track on its own sounded very good. Each guitar track when played with the backing track sounded perfectly acceptable. Nothing at all had stood out in the individual guitar sounds to cause anybody to question the sounds during the recordings. But, when all five guitar tracks were added to the mix, the recording room was clearly evident, and sat very unhappily with the live instruments from the concert. I do not blame any of the recording personnel for this, as they were unfamiliar with the room and had only been given notice of the overdubs four days before the record company said that the tapes needed to be at the CD plant, ready for a Christmas release, but it was a salutary lesson of how the idiosyncratic nature of any room, other than a dead or neutral one, can insidiously build up if used to excess.

A later chapter deals with the subject of vocal rooms, and here again, the unnatural acoustic of many of the typical rooms, particularly some of the older ones, can stand out in an unpleasant way. This is particularly true when listening on headphones, or in many modern control rooms with dead monitoring conditions, when the control room ambience does not swamp the perception of the low level ambience in the recordings. The moral of this story is that if you do have variable acoustic conditions in a smallish room which will be used for many overdubs on the same recording, then *use* whatever variability is available, even if this perhaps means that each individual recorded sound is not at its considered optimum when listened to in isolation. Most of the characteristic sound of the small rooms such as those under discussion here are not frequency aberrations *per se*, but characteristics of the time response. It is not just the frequency content of the room sounds which stand out, but the time-performance of the resonant modes and reflexions which give rise to the idiosyncrasies of the overall decay. As such, with the problems being in the time domain, there is little hope of using the equalisation controls of the mixing console to correct the problem. Any attempts at such 'corrective' equalisation will seriously detract from the frequency balance of the direct sounds, and in many cases, the effect of the medicine will be worse than the illness.

Such room problems can usually only be solved by acoustic changes in the rooms themselves. If small rooms cannot be significantly changed acoustically, then at least the positions of subsequent recordings should be changed. If we are to continue our medical analogies, although the room-related problems may not be *terminal* to a recording, once in the recording system, they do tend to be incurable. In smaller rooms, where the room sound is noticeable, it is of paramount importance to monitor it carefully at the recording stage, and if any sign of a characteristic acoustic build-up becomes evident, steps must be taken to ensure that further recordings on any given song are done with a varied acoustic. This may mean moving the musician(s) and/or the instrument(s), moving the microphone(s), or making use of any acoustic variability. Although changing a microphone type may change the direct/reflective pick-up characteristic, it is not likely to be as effective as the former measures, as the offending time characteristics are quite likely to still find their way into the recordings. Nevertheless, idiosyncratic acoustics do have their place when they are available for effect, so perhaps we should move on to the next chapter; a discussion of live rooms.

Chapter 4

Live rooms – their revolutionary strengths and weaknesses

4.1 A brief history of idiosyncrasy

In the early days of recording, when studios were expected to be able to handle any type of recording which their size would allow, live rooms were more or less unknown. To this day, if the *only* recording space in a studio is a live room, then it is either in a studio which specialises in a certain type of recording, or the room is used as an adjunct to a studio which is mainly concerned with electronic music. Live rooms are distinctly individual in their sound character, and tend to impose themselves quite noticeably upon the recordings made in them. On the other hand, when that specific sound character is *wanted*, then effectively there is no substitute for live rooms. Electronic, or other artificial reverberation simply cannot achieve the same results, a point which will be discussed further in the following chapter on stone rooms.

When, in the late 1960s, many British rock bands began to drift away from the more 'sterile' neutral studios, gravitating towards the ones in which they felt comfortable and in which they could play 'live' in a more familiar sonic environment, a momentum had begun to grow which would radically change the course of studio design. In 1970 the Rolling Stones put the first European 16-track mobile recording truck into action. This was not only intended for the use of live recordings, but also for the recording of bands in their homes, or anywhere else that they felt at ease. It was soon to record the Rolling Stones *Exile on Main Street* album in Keith Richard's rented house, Villa Nelcote, between Villefranche-sur-Mer and Cap Ferrat in the south of France, but another of its earliest uses was the recording of much of Led Zeppelin's fourth album. This album was a landmark, containing such rock classics as 'Stairway to Heaven' and 'When the Levee Breaks', the latter perhaps inspiring a whole generation of recorded drum sounds. As mentioned in the previous chapter, Led Zeppelin had made many recordings in the famous, old, large room at Olympic, in London. For the fourth album they rented Headley Grange in Hampshire, England, and took along the Rolling Stones' mobile, and engineer Andy Johns. The house had some large rooms, but none had been especially treated for recording. Nevertheless, as the house existed in a quiet, country location, the ingress and egress of noise was not too problematical.

4.1.1 From a room to a classic

It seems that 'When the Levee Breaks', with its stunning drum sound for the time, was never planned to be on the album, nor indeed to be recorded at all. In the room in which they were recording, John Bonham was unhappy with the sound of his drum kit, so he asked the roadies to bring another one. When it arrived, they duly set it up in the large hallway, so as not to disturb the recording, and waited for John to try it. At the next available opportunity, John took a break from recording and went out into the hallway to see if he preferred the feel and sound of the new kit. The other members of the band remained in their positions, relaxing, when suddenly a huge sound was heard through their headphones. The sound was from the drum kit in the hallway.

John had failed to close the door when he went out, and the sound from the kit that he was playing was picking up on all the open microphones in the recording room. The hall itself was a wood panelled affair, with a large staircase, high ceiling and a balcony above. It was thus diffusive, reverberant, well supplied with both late and early reflexions, and very much of a sonic character which matched perfectly the style and power of John's drumming. He went into a now famous drum pattern, over which Jimmy Page and John Paul Jones began playing some guitar and bass riffs which they had been working on. Robert Plant picked up on the whole thing, and sung along with some words of an old Memphis Minnie/Kansas City Joe McCoy song. Subsequently, Andy Johns, who had been recording the sounds out of pure interest, reported from the mobile recording truck that they should consider this carefully, as he was hearing a great sound on his monitors.

Such was the birth of this classic rock recording. The story was related to me by Jimmy Page a dozen years or so after the event. I had phoned him whilst I was producing some recordings for Tom Newman (co-producer of Mike Oldfield's *Tubular Bells I* and *II*), as one of the songs being recorded was the one referred to above. The problem which we had was that no matter how many times we listened to the Led Zeppelin version, we could make absolutely no sense of the words in the bridge section. 'Probably that is because they *don't* make sense,' replied Jimmy. Evidently, during Robert's sing-along, there was no bridge lyric which he knew or remembered, so he sung what came into his head. Why I relate this story here in such detail is because it shows, most forcefully, how a *room* inspired an all-time rock classic. It is almost certainly true to say that without the sound of the Headley Grange hallway, Zeppelin's 'When the Levee Breaks' would never have existed.

Had Led Zeppelin been recording in a conventional studio of that time, they could perhaps now have been considered to be one rock classic short of a repertoire. *But*, it must also be remembered that had John been playing a different drum pattern in Headley Grange, or had Jimmy Page and John Paul Jones opted for a different response, then the sound of the drums in the hallway may have been totally inappropriate. This highlights the limitation with live rooms; they can be an inspiration and a unique asset in the creation of sounds, or they can be a totally intrusive nuisance. Furthermore, there is no one live room which will serve all live room purposes.

4.1.2 Limited, or priceless?

Let us suppose that we had a reverberation chamber, similar to the ones used for acoustic power measurements in universities, institutes and research houses. These are constructed to give the broadest achievable spread of modal resonances, in order not to favour any one frequency over another. They *can* produce excellent results as reverberation chambers for musical mixes, but they lack the specific idiosyncratic characteristics that make *special* live rooms special. Let us now consider an analogous situation where guitar manufacturers opt for a totally even spread of resonances in their instruments; to have 'life', but as uniformly spread as possible. All of a sudden, almost all guitars would begin to sound much more alike than they currently do. In general, it is impossible to make a Fender Stratocaster sound like a Gretch Anniversary, or vice versa, because the time domain responses of the resonances and internal reflexions cannot be controlled by the frequency domain effects of an equaliser. Something halfway between a Stratocaster and an Anniversary would be neither one thing nor the other, and whilst it may well be a valid instrument in its own right, it could not replace either of the others when their own special sounds were wanted.

Another analogy exists here in that most professional guitarists have a range of instruments for different purposes: different music, different arrangements, different styles and so forth. No famous guitarist that I know has a 'universal' guitar. Such is the case with live rooms, where a 'universal' room could be used as an acoustic reverberation chamber for the recording and mixing process, but could never be used to substitute for the special sounds of the special rooms. What is more, no large, self-respecting recording studio has banks of identical 'best' electronic reverberation devices, but rather a range of different ones. It is generally recognised that a range of different artificial reverberation is greatly preferable to more of 'the best'. In so many cases, neutrality is quite definitely not what is called for.

I recall speaking to Dave Purple, who was formerly at dbx in Boston, USA. In the 1960s he was an engineer at Chess Records in Chicago, when they had the big EMT 140 reverberation plates. These consisted of spring mounted steel sheets, about 2 m × 1 m, with an electromagnetic drive unit and two contact pick-ups. Dave told me that at Chess they used to heat the room where the plates were kept to about 30°C, then tension the plates. When the room cooled down on removal of the heaters, and if they were lucky and the springs did not snap as the steel contracted, the plates produced a unique sound. More often than not, however, the springs did snap, and the whole process would have to be tediously repeated until it succeeded in its aims. Some of the earliest recordings which inspired me to work in this industry were the Chess recordings of Bo Diddley and Chuck Berry; partly due to their powerful and distinctive music, but partly also for their unusual sounds. Somewhat unusually, for the time, the Rolling Stones went to Chicago specifically to record in the old Chess studios at 2120 South Michigan Avenue, and its unique sounds can still clearly be heard on a number of their older recordings.

The old EMT plates were very inconsistent, and though their German manufacturers guaranteed their performance within a quite respectably tight specification, the sonic differences achievable within that specification were enormous. What Chess were doing would have had the EMT engineers tearing their hair out, but it got Chess the sound which was a part of their fame, so in such circumstances, specifications were meaningless. The irony here is that whilst EMT were trying to make widely usable, relatively neutral plates, Chess were doing all they could to give *their* plates a most definitely un-neutral sound. This situation was a perfect parallel to what was happening in the development of live rooms for studios, where the musicians and engineers were looking for something other than what the room designers were offering them. I recall also, in 1978, when building the Townhouse studios in London, desperately trying to buy a specific EMT 140 plate from Manfred Mann's Workhouse studios. I had done some recordings there for Virgin Records, and had been impressed by the sonic character of one of their plates in particular. I offered them a really ludicrously large amount of money for the plate, but short of buying the whole studio, towards the success of which the plate had no doubt contributed, there was no way that Manfred Mann and Mike Hugg would sell it. In fact at one stage, I actually did come close to closing a deal on the whole studio, which also had a wonderful sounding API mixing console.

In the Silo studios in West London in the early 1980s, they had intended to build a large studio room with a small drum room, connected to the studio via a window. The proposed drum room was about 3.5 m × 2 m, and about 2 m high. This was happening right at the time when the backlash against small, dead, isolated drum rooms was reaching a peak, and once the owners realised that their ideas were out of date, they stopped work on the room. When the studio opened, the proposed drum room remained just a concrete shell, with a window and a heavy door. I recorded some of the best electric rock guitar sounds that I have ever recorded with amplifiers turned up loud in that room. This pleased the owners, as their 'white elephant' had become a famous guitar room. The density of reflexions when the room was saturated with an overdriven valve (tube) amplifier combo was stunning. The 'power' in the sounds, even when at low level in the final mixes, was awesome, yet the problem with this room was that for almost any other purpose it was a waste of space. Neither this room, nor Manfred's plate, nor the Chess plates, nor the Headley Grange entrance hall would have had any significant place outside their use for certain very specific types of music and instrumentation.

In another instance, I was fortunate enough to be one of the engineers on a huge recording of Mahler's Second Symphony for film and television – and what is now one of the CBS *Masterworks* CD series. It was performed by the London Symphony Orchestra, the 300-piece Edinburgh Festival Choir, two female soloists, an organ, and an 'off stage' brass band. It all took place over four days in Ely Cathedral, England, which the conductor, Leonard Bernstein, had specifically chosen for its acoustics and general ambience. This was a masterful choice by a genius of a conductor, but despite the wonderful acoustics of the cathedral for orchestral/choral purposes, it would have been entirely unsuitable for any of the other recordings discussed so far in this chapter. They would simply have been a mess if recorded here. Yet, conversely, perhaps there was no studio that could have hoped to achieve a recording with the sound which we captured in the cathedral.

4.1.3 Difficulties for designers

The Kingsway Hall in London was an example of another great room for the recording of many classics, despite its sloping floor and problems from the noises of underground trains, aeroplanes and occasional heavy traffic. It was an assembly hall, never specifically designed for musical acoustics, yet it far outstripped the performance of any large orchestral studio in London at the time. Unfortunately, I believe that it has now been demolished.

On the other hand, specifically designed concert halls have been criticised for their failure to live up to their expected performance, though such criticism has not always been fair. Concert halls must cater for a wide range of musical performances, so tend to some degree to have to be 'jacks of all trades' rather than 'masters of one', the latter of which is perhaps more the case with Ely Cathedral. To experience Mahler's Second in Ely, or to have heard some Stockhausen in Walthamstow Town Hall would show just how appropriate those venues are, but Stockhausen in Ely Cathedral? … Perhaps not! Unfortunately, though, purpose-built concert halls would be expected to reasonably support both, at least adequately, if not optimally. As yet, you cannot design a general purpose concert hall or recording studio to equal the *specific* performances of some accidentally discovered recording locations for certain individual pieces of music. Neither can you build one live room to equal the performances of Headley Grange, the Silo's concrete room, Manfred Mann's plate, and the old Chess Studios. Brass bands, chamber orchestras, symphony orchestras, choirs, organs, folk music, pan pipes, and a whole range of other instrumentation or musical genres all have their own requirements for optimum acoustic life. The bane about the 'best' live rooms is that they gain their fame by doing one thing exceptionally well, but for much of the rest of the time they lie idle. If one is not careful, they can be something of an under-used investment, which is why many studio owners opt to use whatever limited space they have for rooms of more general usability.

The complexity of the acoustic character of live rooms is almost incomprehensible, and they are often designed on intuition, backed up by a great deal of experience, as opposed to any rules of thumb or the use of computers. Just as no computer has yet analysed the sound of a Stradivarius violin or shown us how to make one, full computer analysis of what is relevant in good live rooms is beyond their present capabilities: the modelling of surface irregularities is not possible, unless the position and shape of every surface irregularity were known in advance, and in greater detail than is realistically possible. Designs are thus usually 'trusted' to designers with experience in such things, but in almost every case, some engineer or other will declare the result 'rubbish' because it fails to do what he or she wants it to do. The lack of a fixed specification for a great room can also lead to other nonsenses.

I once visited a studio with a live room designed by a very well-known acoustician. When I was shown into it by its proud owners, I was shocked. Not only was this a travesty, it was a crime against acoustic decency. I could not understand how a designer of such repute could have done this, and wondered how, if I should later meet the person face-to-face, I could ask how it had occurred, whilst still being tactful. As it happened, I discovered some months later that it was not the designer's fault at all. What had happened was that one of the studio owners did not like the look of the original stone, so ordered the builders to use expensive-looking polished granite slabs. As a result, the place not only looked like a bathroom, but sounded like one as well. Unfortunately, such is the level of acoustic ignorance which all too frequently pervades this industry.

After hearing the reason for the above nonsense, I immediately thought back to a studio which I designed in 1985, funded largely by local government and unemployment reducing schemes. The money had to be spread over a range of departments of the co-operative, and my role as acoustic designer had to pass through a strangely concocted committee. As it is unusual to find good acoustic designers on the unemployment registers, they had to resort to my professional help for this service, but there were trained, unemployed interior designers on the committee. The co-operative asked for a live room, so I designed a room, which had to be relatively inexpensive, with rough slabs of York stone for the walls, and coloured concrete slabs for the floor. However, in my absence the interior designers decided that they could make pretty patterns on the walls with the coloured concrete slabs so instructed the building labourers to reverse the positions. As a result, not only did the room not sound as I had hoped (though they were lucky, as it was not too bad) but they had a devil of a job wheeling pianos and flight cases over a rough flagstone floor. Political correctness, interior decoration and live room acoustics are not always happy bed-fellows. Remember the old saying about a camel being the result of a committee trying to design a racehorse!

4.2 Drawbacks of the containment shell

The question is frequently asked as to why so many great acoustics are found in places not specifically designed to have them, and why so many places which are specifically designed to have good acoustics are frequently not so special. For example, why is the fluke sound of the entrance hall at Headley Grange so difficult to repeat in a studio design? Well, part of the problem was discussed in Chapter 1. Studios tend to be built in isolation shells, which reflect a lot of low frequency energy back into the room. This reflected energy must then be dealt with by absorption, which in turn takes up a lot of space and generally changes the acoustic character of the rooms to a very great degree. The old Kingsway Hall in London was used for so many classical recordings because of its very special acoustic character, but its drawbacks were tolerated in a way which they never would be in a recording studio. Many good takes were ruined by noise, and had to be re-recorded, but because it was *not* a recording studio, this was a hazard of the job. Ironically, a purpose-built recording studio with similar problems would be open to so much argument and litigation that it would not be a commercial proposition, even if it had the magic sounds. People would not tolerate such problems in a building actually being marketed as a professional studio; their expectations would be different.

The Kingsway Hall had windows through which many noises entered, as they also did through the floor and general structure. On the other hand, these were the selfsame escape routes which allowed much of the unwanted sounds to leave, such as the sounds which would cause low frequency build-ups. Structural resonances could, in turn, change the sound character, and remember, once again, that most instruments were developed for sounding their best in the normal spaces of their day. Once we build a sound containment shell and a structurally damped room, we have a new set of starting conditions. However, if we are to operate a studio commercially, without disturbance to or from our neighbours, and offer a controlled and reliable set of conditions for the musicians and recording staff, such a containment shell is an absolute necessity.

Containment shells usually operate by reflecting much of the sound back into the enclosed spaces, and once we bounce the acoustic energy back into the space in a way which is untypical of 'normal' rooms, we then must deal with it in other 'abnormal' ways, the interactions of which are extremely complex. This is especially the case in smaller sized rooms. If we consider again the entrance hall at Headley Grange, it is a relatively lightweight structure which is acoustically coupled via hallways, corridors and lightweight doors into a rabbit warren of other rooms. Effectively, the sound of the hallway is the sound of the whole Grange, and if we were to contain the hallway in an isolation shell and seal all the doors and windows, then we would end up with a room having a very different sound to the one which it actually has. Unfortunately, I have never yet had a client for a studio design who would give me a building the size of a mansion house only to end up with a recording space the size of its hallway, yet this is what it would possibly take to create a sonic replica.

The above fact is borne out by the case of a studio which I once built in a large decaying city. A troublesome area of the city with a very high crime rate had been partially cleared and the residents resettled elsewhere. Only a few businesses remained, but the city was almost broke and could not continue the redevelopment, so had allocated old industrial buildings, very cheaply, to partly public-financed business start-up programmes. I was asked by an established studio company to design a new studio in a building which they had managed to rent very cheaply from the local government, and the building was virtually without any noise-sensitive neighbours. What was more, the nearby traffic was only light, and it was not under the flight path of any airport. Acoustic control shells were constructed in the building, without any containment (isolation) shells. The only isolation wall was the one directly between the control room and the nearest studio room. The low frequency response of the control room was perhaps the most powerful, tight, and musical of any studio that I had ever built.

As so many of the neighbours of hi-fi enthusiasts are all too painfully aware, most domestic buildings have very poor sound isolation, yet it is this fact which makes so much domestic hi-fi sound acceptable in untreated rooms. Whatever acoustic losses the rooms cannot provide by absorption, they provide by transmission (letting the sound pass through). This is one reason why the bass in many homes is more 'true' than in many simple studio control rooms. For isolation purposes, much LF energy must remain trapped in many control rooms by their isolation shells. The performance of the control room referred to in the last paragraph was undoubtedly largely due to the lack of an acoustic containment shell, which is the same reason that so many non-specifically designed recording rooms achieve results that can be hard to mimic when all the design considerations of a specifically designed studio must be addressed.

4.3 Raw materials

There is also another aspect of recording in live rooms which, whilst unrelated to acoustics, is so fundamental in their use that it should be addressed. Reproducing the exact acoustic of Headley Grange would not guarantee a Led Zeppelin drum sound. There were five other very important factors in that equation. John Bonham, his drums, the other members of the band, the song, and Andy Johns. Time and time again I am asked to produce rooms which sound like some given example, but in the case in point, John Bonham was a drummer of legendary power, and he also had the money to afford to buy good drum kits. The engineer, Andy Johns, the brother of another legend, Glyn, had been well taught in recording techniques. He also had the excellent equipment of the Rolling Stones mobile truck at his disposal, and considering his brother's fame, perhaps also had a natural aptitude and a pair of musical ears. He knew where to put microphones, and where *not* to put microphones. He trusted his own ears and decisions, and he had excellent musicians and instruments to record.

Shortly after the death of John Bonham, Jimmy Page and Robert Plant booked the Townhouse studios in London to play back some 24-track tapes which they had been recording. If I remember rightly, this was part of the process of deciding if they could continue with another drummer, or whether Led Zeppelin would die with John Bonham. At the time, we had an assistant engineer, George Chambers, who was very inexperienced, but, as it was only a playback and our other staff were otherwise occupied, we put George on the session. He was petrified during the afternoon before the session, saying that he did not know how to get sounds like Led Zeppelin, and what would he do if they asked him to change a sound or something. He was reassured that there would be no problems, and I distinctly remember an ashen-faced George when the musicians arrived about 7 pm. Around an hour later, he came bursting into the lounge almost speechless with excitement. 'It's incredible, it's incredible' he was saying, 'I just pushed up the faders and it was all there!'

The story took me back to the early 1970s when I found out that I had to record Ben E. King, the former lead singer of the Drifters. I had wondered many times previously how to get vocal sounds such as I had heard on 'Spanish Harlem' and his other big hits. Indeed, I had been faced many times with mediocre musicians who wanted me to create some sort of recording magic, and to transform their voices or instruments into that of some famous artiste. Like George, I had also felt some trepidation before the Ben E. King recording. What microphone should I use; or, was the great vocal sound achieved by some compression technique? Was it some equalisation, or reverberation, or echo? What was more, would my career be ruined when I failed to capture the magic vocal sound? On the day in question, I put up several excellent microphones, and tried to have all options covered; then, surprise surprise, when he arrived, regardless of which microphone I used, I simply had to raise the fader and the magic sound of the voice of Ben E. King was issuing from the monitors in all its splendour.

I hope that I am not labouring this point, but a great live room is not a magic potion in its own right. It can enhance the performance, or even, as we have seen, *inspire* the performance of musicians, but *nothing* can substitute for starting off with good musicians, good instruments, good recording engineers and good music. Good recording equipment is also important, but is generally secondary to the previous items. Good and experienced musicians will react to a good room and play to its strengths. They will not simply sit there playing like robots. The whole recording process is an interactive process, hence the care which must be taken to avoid anything which could make the musicians uncomfortable.

4.4 Practical considerations

Live rooms can inspire performances, both as we have already seen, and as we shall see further in the chapter on stone rooms. Visual aesthetics are also an important contributing factor, but there are some things which just have to be. The acoustical demolition of the rooms by the interior decorators, as we discussed earlier, is a case in point. Some things in live rooms are there because designers have to compensate for other things, such as isolation shell interference or the variability of personnel and equipment which may be placed in the rooms during their use. Remember, they often need to re-create the acoustic of a more usual space within a very unusual shell, and this may in itself require some unusual architecture.

The materials which are used to create these internal acoustics are very important in terms of the overall sound character of the rooms. Wood, plaster, concrete, soft stone, hard stone, metal, glass, ceramics and other materials all have their own characteristic sound qualities. Within the range of current response specification, it may be almost impossible to differentiate between the response plots of rooms of different materials, yet the ear will almost certainly detect instantly a woody, metallic, or 'stoney' sound. In general, all the above materials are suitable for the construction of live rooms, and it is down to the careful choice of the designer to decide which ones are most appropriate for any specific design. The overall sound of the rooms, however, will tend to have the self-evident sound quality associated with each material. Wood is generally warmer sounding than stone, and hard stone is generally brighter sounding than soft stone. Geometry and surface textures also play great parts in the subjective acoustic quality.

4.4.1 Room character differences

I suppose that live rooms should now be split into two groups, reflective rooms and reverberant rooms. The former tend to have short reverberation times, but are characterised by a large number of reflexions which die away quickly. The reverberant rooms tend to have a more diffusive character, with a smooth reverberant tail-off. The reflective 'bright' rooms also often employ relatively flat surfaces, though rarely parallel, and a considerable amount of absorption to prevent a reverberant build-up. The reverberant rooms on the other hand tend to employ more irregular surfaces, and relatively little absorption. It is possible to combine the two techniques, but the tendency here is usually towards rooms which have very strong sonic signatures, and consequently their use becomes more restricted.

The question often seems to be asked as to why flat reflective surfaces in studios usually sound less musical than they do in the rooms of many houses or halls. Well, notwithstanding the isolation shell problem, studios rarely have the space-consuming chimney breasts, staircases, furniture, and other typical domestic artefacts. These things are all very effective in breaking up the regularity of room reflexions, but studio owners usually press designers for every available centimetre of space. It has been my experience that far too many of them are more interested in selling the studio to their clients on the basis of floor space rather than acoustic performance. (Never mind the sound ... look at the size!) Perhaps this is to a large degree the fault of the ignorance of the clients as much as that of the studio owners. There is possibly too much belief today in what can be achieved electronically, and the importance of good acoustics is still not appreciated by a very large portion of studio clients. Of course, those who do know tend to produce better recordings by virtue of having had the luxury of better starting conditions, and leave the mass market wasting huge amounts of time trying to work out exactly which effects processor programme they used in order to get that sound. The answer of course is that good recorded sound usually needs little or no post-processing, unless, that is, the processed sound is the object of the recording.

When people say that a given domestic room has a great acoustic, the usual situation is that on hearing a certain instrument in that room, their attention has been called to the enhancement of *that* instrument in *that* position in *that* room. A whole range of other instruments in the same room, either separately or together, may sound very *un*impressive. I recall once recording an album for a well-known flautist in a cottage. We recorded almost all of the flute recordings in a small toilet, using a Shure SM57 microphone. We wanted a powerful sound, which suited perfectly the style of music, and we were very pleased with the outcome. Nevertheless, this does not mean to say that flutes, *per se*, should be recorded in toilets with SM57s; nor does it mean that studios should have toilet-sized rooms lined with ceramic tiles if they want a good flute sound. Such a room would receive very little use, and anyhow, 95% of the time I would prefer to record a flute with a Schoeps condenser microphone and in a more spaciously ambient environment. However, it only seems to take one famous recording to use a given technique, and a huge proportion of the industry will try to copy the technique, believing that it is *the* way to record that instrument. I hope that I am exaggerating here, but I fear that I am not!

4.5 The general concept

I hope that what this chapter is beginning to get across is the point that there are almost no absolute rights or wrongs in terms of live-room design. Whatever a designer provides, there will always be people coming along who have heard 'better' elsewhere. It is also surprising how one famous recording made in a room will suddenly reverse the opinions of many of the previous critics, who will then flock to the studio for the 'magic' sound. Creating good live rooms is like creating instruments – certainly to the extent that the skill, intuition and experience of the designer and constructors tend to mean more than any textbook rules.

I attempted when writing this to at least give a list of taboos, but every single time that, off the top of my head, I considered something to be absolutely out of the question for wise room design, I found that I could think of an example of a room flouting such regulations from which I have heard great results. Large live rooms are unusual in that, beyond their existence in a somewhat acoustically controlled form, as, for example, the rooms used for selected orchestral recordings, they are normally to be found outside of purpose-designed recording studios. Their use tends to be too sporadic to allocate such a large amount of space to occasional use.

4.5.1 Driving and collecting the rooms

Most smaller live rooms do at least appear to have one thing in common: recording staff must learn how to get the best out of each one individually. The modal nature of such rooms defies any reasonable analysis: the complexity is incredible. The positions of sound sources within the rooms can have a dramatic effect on determining which modes are driven, and which receive less energy. An amplifier facing directly towards a wall will, in all probability, drive the axial modes very strongly. However, precisely to what degree they will be driven will depend upon such things as the distance between the amplifier and the facing wall, or whether the loudspeaker is in a closed box, or is open-backed.

Positioning the amplifier (or at least its loudspeaker) on an anti-node, where the modal pressure is at a peak, will add energy to the mode, and resonate strongly at the natural frequencies of the mode. Placed on a node, where the pressure is minimal, that mode will not be driven, and its normally strong character in the overall room sound will be overpowered by other modes. Positions in between will produce sounds in between. If the cabinet is open-backed, it will act as a doublet (or figure-of-eight source), radiating backwards and forwards, but very little towards the sides. A closed-back cabinet at lower frequencies, say between 300 and 400 Hz, will, if facing down the length of a room, drive the low frequency axial modes of all three room axes – that is, floor to ceiling, side wall to side wall, and front wall to back wall. With an open back, however, only one axial mode will be driven, the axis along which the loudspeaker is facing. Figure 4.1 shows the typical radiating pattern of open- and closed-backed cabinets.

Microphone positioning in such rooms is also critical. A microphone placed at a nodal point of a mode will not respond to that mode, as it is at the point of minimal pressure variation. Conversely, at an anti-node (a point of peak pressure), the response may be overpowering. Microphone position can be used not only to balance the relative quantities of direct and reverberant sound, but also to minimise or maximise the effect of some of the room modes. Furthermore, more than one microphone can be used if the desired room sound is only achievable in a position where there is too little direct sound. Another parallel to the ability of amplifiers to drive a room is the ability of the microphones to collect it. Variable pattern microphones can produce greatly different results when switched between cardioid, figure-of-eight, omni-directional, hyper-cardioid, or whatever other patterns are available. What is more, certain desirable characteristics of a room may only be heard to their full effect via certain specific types of microphones.

If the amplifiers are set at an angle to the walls, they will tend to drive more of the tangential modes, which travel around four of the surfaces of the room. If the amplifiers are then also angled away from the vertical, they will tend to drive the more numerous, but weaker, oblique modes. At least this

Figure 4.1 Comparison of open- and closed-back loudspeakers in the way that they drive the room modes

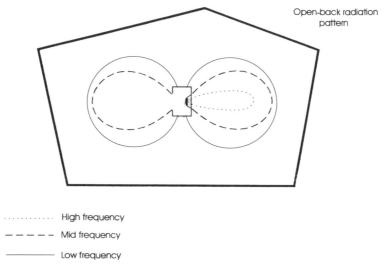

Open-back radiation pattern

··········· High frequency

− − − − − Mid frequency

——————— Low frequency

(a) High frequencies propagate in a forward direction, but low and mid frequencies largely radiate in a figure-of-eight pattern

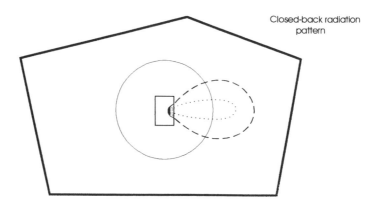

Closed-back radiation pattern

(b) As with (a), the high frequencies still show a forward directivity, but, in this case, the mid frequencies also radiate only in a largely forward direction, and the low frequency radiation pattern becomes omnidirectional

will tend to be the case for the frequencies which radiate directionally. Precisely the same principle applies to the directions which the microphones face in respect of their ability to *collect* the characteristic modes of the room. If more than one microphone is used, switching their phase can also have some very interesting effects. Given their positional differences, the exact distance which they are apart will determine which frequencies arrive *in* phase and which arrive *out* of phase, but there is no absolute in- or out-of-phase condition here, except at very low frequency in small rooms. A pair of congas which sound great in one live room, may not respond so well in another. Conversely another set of congas in the first room may also fail to respond as well. But, there again, in another position in the room perhaps they will. It could all depend on the resonances within the congas, and if they matched the room dimensions. The whole thing depends on the distribution of energy within the modes, the harmonic structure of the instruments, and the way the wavelengths relate to the coupling of the room modes.

4.6 The learning curve

Hopefully, from all this discussion, it will by now be clear that live rooms are musical instruments, and share many of the characteristics of other musical instruments. The proliferation of freelance recording engineers and producers, travelling from studio to studio, often leaves them having to deal with a strange live room in a somewhat similar way that musicians have to deal with strange instruments. Musicians, in fact, generally do not like to have to play strange instruments. Their instruments are very personal, and over a period of time, they get used to the feel of them and learn how to get the best from them. The frequent travelling between studios does not always allow freelance engineers the time to learn how to get the best from many live rooms, which, in the hands of a permanent and experienced studio staff, would be capable of producing some very interesting sounds. We could even draw a parallel with sports, such as Formula One motor racing. All drivers entering a Grand Prix race have a great deal of experience, yet if they were all to swap cars at the beginning of a race, perhaps very few would even *finish* the race. Racing without having learned the individual feel of the strange cars would lead to a chaotic and dangerous race. In fact if such a race was proposed, I am sure that many drivers would refuse to take part. However, without the life-threatening risks in the studio business, people do sometimes launch themselves on such misguided pursuits, then wonder why the results do not come up to expectation. No matter how much general experience an engineer may have, learning about live rooms is a very individual process. It takes time to get to know them, at least if their *full* potential is to be realised.

Live rooms are great things to have as adjuncts to other recording facilities, but they are a dangerous proposal if they form the only available recording space, unless the studio is specialising in the provision of a specific facility and the recording staff know the characteristics of the room very well indeed. These rooms have grown in popularity as it has become apparent that electronic simulation of many of their desirable characteristics is way beyond present capabilities. They have also been able to provide individual studios with something unique to each one, and this point is of growing importance in an industry where the same electronic effects with the same computer programmes are becoming standardised the world over. If a certain live room sound is wanted, it may well bring work to the studios which possess it, as the option to go elsewhere does not exist. But be warned, they can bite! The design *and* use of these rooms are art-forms in their own right, and very specialised ones at that.

Chapter 5

Stone rooms

Live rooms have been constructed from wood, metal, glass, ceramics, brick, concrete and many other reflective materials. Undoubtedly, the materials of construction affect the timbre of the resulting sound and it is very difficult, if not impossible, to make one material sound like another. I frequently use wood in live areas, but largely due to the success of certain early rooms, I seem to be asked to build stone rooms more frequently than others. Stone is unique amongst the other materials in that its surface, in a readily available state, is far less regular and hence more diffusive than the other materials. Hard stones sound different to soft stones, just as hardwoods sound different to softwoods. No one live room can be all things to all people, but where a specific live room is required, as opposed to a live area *within* a room, stone does seem to be a particular favourite for many. The widespread use of stone rooms is a relatively recent phenomenon and, as with so many other aspects of the recording industry, its origin and acceptance can be traced back through some quite unpredictable chains of events. Certainly the evolutionary path of the many rooms which I have built could neither have been foreseen nor controlled.

5.1 Evolution

Twenty-four track tape recorders became generally available in early 1973, bringing what was an unprecedented luxury of being able to record drum kits using multiple microphones on a one microphone to one track basis. Seven years earlier, anything more than two or three microphones and one track of the tape recorder for a drum kit was seen as either wanton extravagance, or the actions of an ostentatious 'prima donna' engineer. Almost inevitably, the one microphone to one track recording technique led to experimentation in the recording of drum kits. Equalisation of the individual drums in the kit was a novelty that was much pursued, but the desired signal processing was seen to be being made less simple or effective by overspill from adjacent drums in the kit picking up on the other microphones. The answer seemed to lie in increased acoustic separation, which led in turn to studio designers being asked to produce isolation booths which not only separated the drum kit from the rest of the instrumentation, but also enabled a new degree of separation between the individual drums within a kit. The mid-1970s witnessed the industry experiencing the aberration of the use of very dead drum booths, which were later to be largely relegated to use for the storing of flight cases or microphone stands.

The highly damped drum booths had enabled the period of experimentation with separation to run its course. Achieving separation was undoubtedly facilitated by the use of these booths, but unfortunately, they posed two major problems. Firstly, once the novelty of individually equalised and processed kits had begun to wear off, the essential ambient 'glue' which held a kit together was conspicuous by its absence. Secondly, and probably even more importantly, most drummers did not enjoy playing in these booths. An uncomfortable drummer can rarely produce an inspired performance, and as an excellent recording of an uninspired performance would rarely last the test of time, it soon became apparent that the human requirements of the drummers required far more attention paying to them than had previously been allocated: at least, that is, if inspired performances were required. Drummers soon began moving back into the main studio areas, often reverting to the older practice of being shielded behind acoustic screens, but the re-recognition of the importance of ambient 'glue' to hold a kit together was beginning to signal the end of the dead isolation booths for the drums.

My personal recollection of the advent of the stone rooms began in early 1978 when I was discussing the designs for Townhouse One, in London, with Tom Hidley and Mick Glossop (who had been scheduled to be the chief engineer when the Townhouse opened). I remember pushing hard for at least one of the four isolation booths to be given a live acoustic, especially as I had envisaged (wrongly) that once open, the Townhouse would have relied to a much greater degree on 'session' work with strings and brass, rather than the 'lockout' rock music bookings which became its bread and butter. I managed to gain a consensus that we should have a wooden floored, wood/glass/mirrored walled room at the back of the studio. Mid-way through that year, Richard Branson decreed that a second recording studio was to be constructed in place of the originally proposed rehearsal room. Many aspects of the two control rooms were to be similar, but this time, in the second studio, there was to be scope for experimentation.

I proposed a stone room, expanding on the theme of the studio area at The Manor Studio in Oxfordshire, England, completed some three years earlier, but as soon as I voiced my intentions, I ran headlong into Tom Hidley's design performance guarantees. Tom was trading very largely on a promise of *provable* performance, but as my proposals were deemed 'unpredictable', Tom announced that he could not be associated with such a room. However, I had decided what I wanted, and had agreed with Tom's sub-contracted builders that they could construct the room as planned. I offered to sign a waiver exonerating Tom from any repercussions if the room failed to perform, but even in my capacity as technical director of all the Virgin Studios, he still would not go along with the room. Not until Richard Branson, himself, signed the waiver would Tom give the green light for his construction team to go ahead. Incidentally, none of this should be seen as being in any way a criticism of Tom Hidley; times were *very* different then, and I knew that I was sticking my neck out.

Following from The Manor, the Townhouse Two room was built from Oxfordshire sandstone. The floor was Clipsham stone, a blue stone with pink veins which I had first encountered in Mike Oldfield's kitchen whilst mixing his 'Boxed' compilation album. The theatre floodlights and lighting bar were from the original Manor Studio, prior to its reconstruction, and the wooden ceiling was as per the control room. The playback monitors were purchased from Majestic Studios, where I had designed and installed them in 1970. I suppose that if I *was* sticking my neck out, then I was doing so quite cautiously, as every individual component of the room was already well known to me. Only the whole ensemble was new, but the question still remained as to how it would be received.

The initial response was not good, largely because most engineers were unaccustomed to such rooms, particularly the younger engineers. One of the assistant engineers allocated to that studio was Hugh Padgham. Studio Two was the first in the country to install a large SSL computerised console, so the Townhouse had to allocate specialised helpers to engineers coming from outside. Hugh initially kicked up quite a fuss about resonance problems when recording Hammond organs and acoustic guitars, for which the room was *never* intended to be used, and he also complained about isolation from the control room – a valid point, as the room was loud by conventional standards. Tom Hidley was counting his blessings that he had demanded a waiver, and Richard Branson was ordering me to rebuild the room on more conventional lines. Nevertheless, *I* still believed in the room, and initiated delaying tactics to stall its demise. I installed a third glass door system to improve isolation to the control room, then conveniently disappeared with Mike Oldfield for some time. If ever there was a person who I saved from shooting himself in the foot it was Hugh Padgham. Shortly after the aforementioned occurrences, Hugh was working on the sessions which were to become the Phil Collins' *Face Value* album. The drum sound on 'In the Air Tonight' was eventually to do no harm whatsoever to Phil Collins, the Townhouse, *or* Hugh Padgham – or me for that matter!

Shortly after, I built a larger version at The Manor, but the advent of digital reverberators and room simulators, especially the programmable variety, seemed initially to many people to sound the death knell for the live rooms. As the strengths and weaknesses of the acoustic and electronic approaches began to become more widely appreciated, it was soon understood that each had their place. By the mid-1980s, the live rooms had returned with a vengeance. Now that it is realised that many benefits of the acoustic approach are unattainable by electronic means, it is interesting, when looking back, to see how the thirty-odd live rooms which I have built since the Townhouse would probably never have come about but for the chain of events mentioned earlier. It was a chain of events which depended equally as much on fads, fashion, technology and chance, as on experience, planning, forethought and will. The stone room in Townhouse Two is shown in Fig. 5.1.

5.2 Construction options

One thing which most certainly can be said of stone rooms is that they are *all* different. In these days of pre-programmed instruments and factory set effects programmes, stone rooms add an extra degree of variation. From a personal point of view, I like to vary the designs, not only for overall variety,

Figure 5.1 Townhouse Two, London (1978)

but also so that each studio owner has a genuine 'first edition'; something unique, which, as experience of its performance is gained, can supply a sound unattainable elsewhere. As well as sizes and shapes, the stone which I use has also changed as the years have passed. The early rooms were Oxfordshire sandstone, which was a little on the soft side and slightly crumbly. Consequently, a thin coat of polyurethane varnish was applied to reduce the dust problem. Subsequently, Purbeck and York stones were used, and later still, Spanish and Portuguese granites.

Granites allow a greater degree of variation in acoustics, as, being much harder, they have less tendency to shed dust. Once a PVA adhesive has been added to bind the cement, the choice of varnishing the stone or leaving it bare introduces an interestingly different variable. Varnish noticeably softens the sound when compared to the natural granite, whereas with the softer stones the varnish treatment is less readily noticeable. The high concentration of PVA adhesive in the cement, together with expanded metal backings, allows the cement between the stones to be cut back, and thus exposes in a great deal of relief the outline of each individual stone. This technique was originally developed for the large drum room in Blackwing, London, initially as a cosmetic measure to produce a castle-like atmosphere, but its acoustic advantages also soon become clearly understood. The deep crevices between each stone gave a much more diffusive sound-field, especially as the stones in that particular studio were laid in a highly random manner; again initially for cosmetic effect. See Fig. 5.2.

Figure 5.2 Splendid (Blackwing), London (1988)

One major problem which was solved in the design of the first Blackwing room was how to stop the low frequency build-up, which had previously been problematical in rooms of that size and over. Blackwing was a large room by normal standards, 24 ft × 16 ft × 10 ft (8 m × 5 m × 3 m) after the internal finishes. The shell of the room was much higher, as the space had previously been used as a rehearsal room with a mineral-wool suspended ceiling. There was also a large amount of mineral wool, a metre deep, over this ceiling, and the aversion of everybody concerned towards being rained on by such unpleasant substances during their removal concentrated minds wonderfully in the pursuit of alternative solutions. I eventually elected to leave the ceiling in position, especially as, unusually for such ceilings, it sloped. A coat of bonding plaster, roughly applied, took advantage of that slope to produce a non-parallel, reflecting surface opposing the concrete floor, with the roughness of the plaster helping the high frequency scattering.

The intention was to reflect mid and high frequencies back into the room, whilst allowing the low frequencies to penetrate. For the low frequencies to be reflected back into the room, they would have to penetrate the ceiling, suffer absorption by the metre or so of mineral wool overlay, reflect from the structural ceilings (two of them), and return through the same obstacle course. Obviously, such a path would introduce severe attenuation of those low frequencies, so the ceiling would provide an escape path for them, and act as a high pass filter in one of the three main axes of the room (the vertical axis). On the day the room was completed, in all respects but for the plastering of the ceiling, I can only say that I was glad that it was not one of my first ventures into such areas, or suicide may have resulted. Once the construction tools and materials had been removed, absolutely everybody who came into the building entered the room, clapped their hands, shouted 'one-two' and shook their head with that knowing look of informed disapproval. The room was bright, of that there was no doubt, as there were many reflexions from the hard irregular walls, but reverberation was all but absent to any significant degree.

Fortunately, I had built a dozen or more relatively similar rooms prior to this one. True, this was the first with the highly random stonework and the deeply cut back cement, but even given the new design of ceiling, it *had* to work. It was controlled by the same laws of physics which had served me so well in the past. Nonetheless, it seemed that there was just one layer of plaster between me and acoustic oblivion. The opinions of the observers were virtually unanimous that it would be disappointing, but one stroke of fortune was that the studio owner, Eric Radcliffe, the producer of Yazoo's *Upstairs at Eric's* album, had a background of science degrees and had completed a three year post-graduate course in laser physics at Imperial College before his record production activities really took off and led him away from University. At least he did have an understanding of wave behaviour, and kept faith with my proposals to the end.

The plasterer arrived to do his deed, with the studio owner and myself popping in at regular intervals to check on progress. There were some interesting academic conversations taking place, as it transpired that the plasterer had a degree in sociology, and his brother, who was one of the granite block layers, had a BSc in physics. It really is remarkable where some of these people end up! When the plastering was complete and the trestles and buckets had been removed, we entered once again to watch the plaster dry ... seriously! It was a total revelation; minute by minute, once the plaster had begun to set, the room came to life. With the plaster fully hardened the next day, the room delivered all that had been promised, and more besides. The owner was so pleased that he asked for a second room to be built in his other studio, utilising the same principles but with only about 60% of the room size. They are very different rooms in subjective terms, yet there is an unmistakable family resemblance.

5.3 Live versus electronic reverberation

I still consider the first of the Blackwing rooms to be something special, though I cannot say 'better' or 'worse' than any other. Each stone room is unique. The knack of using them is to play to their individual strengths and to avoid their individual weaknesses. They add a degree of uniqueness to a studio which is simply not yet available with electronic devices; programmable, or not. There are some subliminal reasons for this, but there are also some very hard engineering reasons. Although low-level effects in the tails of digital reverberator responses are generally very low indeed, we do often seem to detect them by their absence in natural reverberation tails. Some aspects of these decay differences have been difficult to measure, as we do not have analytical equipment even approaching the discriminative ability of the human ear/brain combination, but what is more, many of the arguments about just what is, or is not, audible has been based on research into hearing thresholds relating to language and intelligibility. There have been many cases of accident victims whose injuries have resulted in severe impairment of their ability to communicate verbally, yet their appreciation of music has been unimpaired, implying that the areas of the brain responsible for the perception of speech and music are quite separate. Much more research is still required into these differences. Indeed, there is no integrated signal reaching the brain which resembles an analogue of the ear-drum motion. The ear presents the brain with many component factors of the 'sound' and it is only by way of a massive signal processing exercise by the brain that we hear what we hear.

It is the degree of these subtleties which still confounds the manufacturers of digital reverberators. I remember the late Michael Gerzon talking to me about the then current state of electronic reverberators in 1990. Michael was co-inventor of the SoundField microphone and the main developer of the Ambisonics surround sound system, and had much experience in the world of sound-field perception. He said that the state-of-the-art, digital reverberation unit, in electrical terms, represented something in the order of a *ten thousand* pole filter. The complexity of the inter-reaction of the sound-field within a moderately sized live room would be in the order of a *one hundred thousand million* pole filter. Even if the current rate of acceleration of electronic development were to be maintained, it would be forty years or so before a room could truly be simulated electronically; and even then, at what cost, and with what further restrictions?

A room simulator may well go a long way towards reconstructing the reflexion patterns for a sound emanating from *one point* in that phantom 'room'. However, the nub of the issue is that in real life, a band, or a drum kit, does not inhabit one point in space in the room. Different instruments, or different parts of *one* instrument, occupy different spaces in the room. Sounds are generated from very many different positions in the room, some at nodes, some at anti-nodes, and others at many points in between. All excite different room resonances to differing degrees and all produce reflexions in different directions. This subject was discussed in the previous chapter, and depicted in Fig. 4.1. In other words, the room behaves differently towards the snare drum than it does to a floor tom in the same kit. With current digital reverberation, all the instruments, and indeed all the individual *parts* of all the instruments, are injected into the phantom 'room' from, at most, only a few points in the theoretical space. All injections into the same space are driving a similar series of resonances, all are equally distanced from the rooms nodes and anti-nodes. Such occurrences do not exist in nature.

Acoustic reverberation 'chambers', when driven by a large stereo pair of loudspeakers, can overcome this problem to some degree, but what any 'after the event' processing system lacks is the inter-reaction between say, a drum kit, a live room in which it may be played, *and* a drummer. The drums excite the room, and the room resonances, in turn, re-excite, inter-react and modify the resonances of the kit. These processes undertake reiteration until energy levels fall below perceivable thresholds. The instrument, the room, and the musician are inextricably linked; they behave as one complex instrument. Physical separation of the playing and the addition of reverberation break this very necessary unification. The room resonances modify the *feel* of a drum kit to the drummer, and that drummer will also perceive the room effects via any headphones, bone conduction, and general tactile sensations. The room will modify the musicians performance; and this *cannot* be accomplished after the event. Performances are unique events in time. It was on these grounds that the drummers rebelled against the mid-1970s, dead, high-separation drum booths which were then in vogue. I doubt that electronic simulation will ever have an answer for the human performance inter-reaction problem, as no subsequently applied artificial reverberation can acoustically feed its effects back into the feel of the instruments themselves. Only artificial reverberation in the room itself, at the time of playing, could achieve that.

Building stone rooms, or live rooms in general, is a very long way from the acoustic discipline of control room design. Control room design usually seeks a neutrality in which the sound of the room is perceived to as small an extent as possible. For a control room to add any characteristic sound of its own is greatly frowned upon. Conversely, if a live room *sounds* good, then it *is* good. It is possible to walk away from a completed stone room with a degree of satisfaction, pleasure, excitement, and a sense of achievement in having created something different, which can never be attained from control room design. The only drawback to stone rooms would appear to be that they take up more space than a digital room simulation unit, they cannot be taken from studio to studio, and they cannot readily be traded-in or sold-off. People seem to expect to have all of their equipment encapsulated in boxes these days, and from this point of view, the stone rooms can comply ... other than for the fact that they are not rack-mountable, they are somewhat large, and tend to weigh in the order of twenty tons!

5.4 The 20% rule

The story of the ceiling at Blackwing, in London, highlights a point of general significance in terms of the percentages of room surfaces which are needed to create any significant effect. With the mineral wool ceiling it was difficult for the room to achieve any reverberation, as all of the energy in the oblique modes, which passes in a chain around all the surfaces of the room, would be absorbed upon coming into contact with the ceiling, and could thus never become resonant. Much of the energy of reverberation tends to be in these irregular and multitudinous oblique modes. In the 8 m × 5 m × 3 m room, the total surface area is about 160 m². The ceiling (8 m × 5 m) has an area of 40 m², which is about 25% of the total.

A reverberant room generally needs to have reflective material on all of its surfaces, and usually only about 20% of the total surface area needs to be made absorbent to effectively kill the reverberation. On the other hand, in a very dead room, introducing about 20% of reflective surfaces will usually begin to bring the room to life. Equally, a room with troublesome modal problems will usually experience a significant reduction of those problems by the covering of about 20% of its total surface area with diffusers. If one wall creates a problem, the covering of about 20% of that wall with diffusers will usually render the wall more neutral, but in this case, the diffusion should be reasonably evenly distributed over the surface to be treated.

At Blackwing, it was absolutely fascinating to listen to the plaster dry, or rather to listen to the effect of its hardening on the room acoustics. The wet plaster was not very reflective, so upon initial application it did not significantly change the room, but once the hardening process began, after a few hours, it was possible to witness an empty room changing very noticeably in its character, in a way which, in my experience, was almost unique. Without any physical disturbance or any abrupt changes, the room was 'morphing' from a bright, reflective room, to a highly reverberant one. The luxury of such experiences can provide much insight into the acoustical characteristics of rooms of this nature.

5.5 Reverberant rooms and bright rooms – reflexion and diffusion

The terms brightness and reverberation often tend to get confused by the loose use of the term 'live room' when studio acoustics are discussed with many recording personnel. Stone rooms can be produced to either bright or reverberant specifications, and the flexibility is such that rooms looking very similar may sound very different. Figures 5.3–5.6 show rooms of very different acoustic characteristics. All four are built with Iberian granite, and an explanation of the different construction techniques used in each case will help to give an understanding of the respective processes at work.

5.5.1 Pseudo-reverberation

When I have been referring to the reverberation in these rooms, it is not reverberation in its true acoustic sense. Reverberation refers to a totally diffusive sound-field, where its intensity and character is the same throughout the room. This cannot occur in small rooms, as the existence of modal resonances and discrete reflexions will always ensure that places of different character can be found within the room. Even the absolute decay time of all reflective and resonant energy can be position dependent, but in general, the term reverberation, as understood amongst most recording engineers, is perhaps the most widely understood term I can use in these descriptions. Certainly in the names of many programmes in digital effects processors, the term 'reverb' is now so universal that, except in academic acoustic circles, it would be like trying to swim up a waterfall to be too pedantic about its accurate use at this stage of the development of the recording industry. Anyhow, bearing the terminological inexactitude in mind, let us look at some different 'reverberant' room designs.

The room in Fig. 5.3 (the former Planta Sonica studio in Vigo, Spain) was about 5 m × 4 m × 3 m high, and was built using a facing of granite blocks, about 10 cm in thickness. The stones were bonded by cement to a studwork wall, which was covered in various sheet materials such as plasterboard, chipboard and insulation board. The blocks had reasonably irregular surfaces, but were all laid flat against the wall. After the cement behind and between the blocks had dried, the gaps were pointed with a trowel, to smooth over the cement and bring it more or less level with the face of the blocks. The resulting wall was hard and relatively flat, but the irregularities were sufficiently large to be somewhat diffusive at frequencies above about 3 kHz. None of the walls were parallel, which helped avoid the build-up of regular patterns in the axial modes. The ceiling was heavy, and of quite a solid structure behind the plaster. A fabric panel at one end of the ceiling covered the entrance to a low frequency absorber system, which helped reduce excessive low frequency build-up. The

Figure 5.3 Planta Sonica, Vigo, Spain (1987)

room had two sliding glass doors, each about 1 m 80 cm wide and 2 m high, one leading to the control room, and the other to the main recording space.

The Planta Sonica room in Vigo, Spain, produced some excellent recordings of drums, electric guitars, acoustic instruments, and especially the traditional Celtic bagpipes, which are very popular in Galicia, the Celtic province of Spain where Vigo is situated. The room decay was smooth, without excessive low frequencies, and spacially very rich sounding. The empty reverberation time was about 3 seconds, but of course the empty state is not really relevant in such cases because it will not be used empty, except perhaps as a reverberation chamber during mixdown. In such cases, a loudspeaker/amplifier is fed from an auxiliary output of the mixing console, and a stereo (usually) pair of microphones picks up the room sound for addition to the mix. However, in normal use, the influx of people and equipment can have a great impact on the measured, empty performance, as they tend to be absorbent and can occupy a significant portion of the total room volume.

The rooms in Figs 5.4 and 5.5 (Discosette 3 and Regiestudio, Portugal) are built in very similar manners to each other. Both have layers of granite covering the same sort of stud wall structure that the Planta Sonica room had, but they are built more on the lines of Blackwing (see Fig. 5.2). Both also have the granite blocks laid in a more three-dimensional configuration, with many of the blocks protruding from the walls. They have somewhat similar types of ceiling structures, but are much smaller in size, the Discosette room having a total surface area of about 90 m^2, and the Regiestudio room about 60 m^2. The latter room also differs in being made from granite blocks that have only a quarter of the surface area of those in the other rooms.

The Discosette 3 room (see Fig. 5.4) is a strongly reverberant room, but lacks the powerful reflexion patterns of the old Planta Sonica room. The deep

Figure 5.4 Discosette 3, Lisbon, Portugal (1991)

cutting back of the cement between the blocks creates a series of randomly sized pits which render the surface much more diffusive, a property also enhanced by the protrusion of many of the blocks. Due to the non-parallel nature of the surfaces of the room, most of the energy will be concentrated in the tangential and oblique modes, and thus the depth of the pits and protrusions is effectively increased, because the modes will be at varying angles

Figure 5.5 Regiestudio, Amadora, Portugal (1992)

to the wall surfaces (see Fig. 5.7). The effect is that the walls are diffusive down to much lower frequencies than the walls of the Planta Sonica room and there is, therefore, more diffusive energy in the decay tail.

The reduction in the spread of reflective energy renders the clearly defined reflexions and resonances that do exist to be more readily noticeable, even though the total energy which they contain is a lower proportion than that of the Planta Sonica room. Dependent upon instrument and microphone position, such rooms as the one being discussed here can have either more or less predominant character than rooms of the Planta Sonica type. The more diffusive room does not produce the same haunting wail of bagpipes, but can produce some very powerful sounds from congas, and rich enhancement of saxophones or woodwind. These two types of rooms are not really interchangeable; they are sonically very different.

5.5.2 Bright rooms

Let us now move on to consider the Regiestudio room (see Fig. 5.5) in more detail. This is a room with a total surface area of about 60 m^2, about 4 m^2 of which are non-parallel glass surfaces. There is also 2 m^2 of flat, wood-panelled door, and around 8–10 m^2 each of wooden floor and sloping, highly irregular, plastered ceiling. The remaining surfaces of the walls, which form the majority of the surface of the room, are of small granite blocks, each having a face area of 80 to 100 cm^2. The cement has been deeply cut back in the gaps, producing a series of irregular cavities, but unlike the rooms in Figs 5.1 to 5.4, the ratio of the surface area of the granite to the surface area of the pits is much less; in fact only around 20% of that in the other rooms.

The effect of this is to produce far fewer specular reflexions because there are fewer flat surfaces, at least not acoustically so, until much lower frequencies. Any wave striking the wall surface will be reflected differently from the stones and pits, but let us consider the case of what happens at 500 Hz, where the wavelength is around 60 cm. If the pits average 8 cm in depth, then a reflexion travelling from the face of the stones will travel about 16 cm less than when entering and being reflected from the back of a pit. This will create about a quarter-wavelength phase shift on a single bounce, and the irregular shape of the stones and pits will tend to scatter the wavefront as it reflects from the wall. At least down to around 500 Hz, such a room surface becomes very diffusive until specular reflexions begin again when the dimensions of the stone faces become proportionate to wavelength, say above 5 kHz. However, at these frequencies and above, the irregularity of the stone surfaces themselves begin to become diffusive, so the walls begin to scatter from around 500 Hz upwards. There will also be a considerable degree of diffraction from the edges of the stone blocks, which will also add to the diffusion.

The Regiestudio room is very bright, emphasising well the harmonics of plucked string instruments, and adding richness to flutes and woodwind. Somewhat surprisingly perhaps, the reverberation time is much shorter than one would tend to expect from looking at the photograph. In this type of room, which is also very small, the energy passes rapidly from surface to surface. As the surfaces are so diffusive, the scattering of the modal energy is very wide. The speed of sound is dependent on temperature, and as almost all recording studios will be kept at relatively similar working temperatures, the speed of sound can be considered to be the same in all of them. Thus in a room such as this, the number of times that a sound wave will strike a wall surface on its travels around a room are many times greater than would be the case in, for example, the Blackwing room (Fig. 5.2). Each impact with a wall surface, especially at a near grazing angle as opposed to a 90° impact, will rob energy from the reflected wave, either by absorption or transmission, or by the energy losses due to the interaction of the diffusive elements. Consequently, in two rooms of any given surface material and construction, the small room will have the lower reverberation time because more surface contacts will take place in any given period of time. They will also have a higher initial reflexion density. The Regiestudio room produces a brightness and thickness to the recorded sound, but it falls off within about one second.

The Shambles room (see Fig. 5.6) shows a further variation on the theme, and is a room with a character somewhere between the ones shown in Figs 5.3 to 5.5. It is the only recording space of a small studio, so has to be slightly more flexible than the 'specific' rooms shown in the other photographs. When empty, it has a reverberation time of just over 2 seconds, but this can readily be reduced by the insertion of lightweight absorbent 'pillars' containing a fibrous filling, especially when they are positioned some way out from the corners. As we discussed in Chapter 2, fibrous absorbers are velocity dependent, so should not be placed too near to a reflective surface or their effect will be reduced.

From the descriptions of the rooms shown in Figs 5.1 to 5.6, it can be appreciated that the cutting back of the cement to form pits between the stone blocks both increases the diffusion and lowers the decay time. Figures 5.7 and 5.8 will help to show the mechanisms which create these effects. Figure

Figure 5.6 Shambles, Marlow, England (1989)

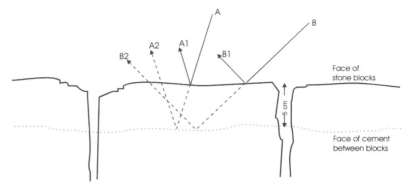

Figure 5.7 Effect of incidence angle on the reflexions from the pits and blocks. From a nominal depth of cavity of 5 cm, incident wave A, when it reflects from the back of the pits, will travel about 10 cm further before returning to the room, as compared to when it reflects from the face of the blocks. Incident wave B, striking at a shallower angle, will show an even greater pathlength difference between its reflexions from the back of the pits and its reflexions from the faces of the blocks

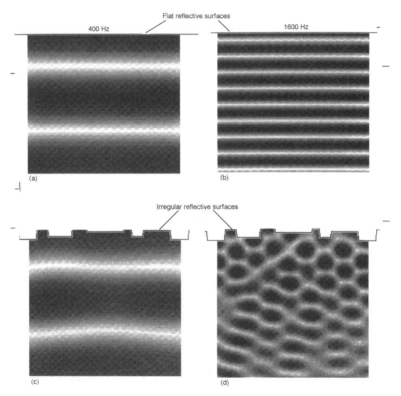

Figure 5.8 Effect of surface irregularities on reflexion patterns. Figures (a) and (b) show the interference patterns of 400 Hz and 1600 Hz plane waves when reflecting from a flat surface. Figures (c) and (d) are again 400 Hz and 1600 Hz plane waves, but this time the interference fields are those produced by relexions from an irregular surface, such as a rough stone wall. In (c) the 400 Hz pattern is little changed from (a), but at higher frequencies, as shown in (d), where the size of the irregularities becomes significant in proportion to the wavelength, the reflexion pattern becomes fragmented, and loses energy more rapidly after reflexion

5.7 shows the way in which the reflexion paths differ from the front of the blocks and the back of the pits. It can also be seen that for angles of incidence other than 90° the disruptive effect will be greater as the pathlength differences increase. The frequency dependence of the effect of the irregularities is shown in Fig. 5.8. At 400 Hz, there is little difference in the interference patterns created when a plane wave strikes either the flat surface or the irregular surface at a 90° incidence angle. At 1600 Hz, however, the reflected wave is well broken by the 10 cm depth of surface irregularity. It can also be seen that the energy in the pattern reduces significantly more rapidly further away from the wall, which is a result of the diffusive effects.

In normal situations, the effect is even more pronounced than that shown in Fig. 5.8 because the rooms which employ such surfaces usually have non-parallel surfaces. This tends to cause more of the sound waves to strike the wall surfaces at angles of incidence other than 90°, where the pathlength differences caused by the irregularities will be greater. The two primary effects of this are that the description of the interference field will extend to lower frequencies, and the energy losses after reflection and diffraction will be greater.

In Section 2.4.1, we were looking at the question 'What is parallel?', and saw that the degree to which a pair of surfaces were *acoustically* parallel was very frequency dependent. Somewhat similarly, Fig. 5.8 shows the highly frequency-dependent nature of the effect of surface irregularities. At 1600 Hz, the effect of the surface irregularities can clearly be seen (and heard). At 400 Hz, the effect of the surface irregularities is only minimal in comparison to the interference pattern produced by an absolutely flat wall. Down at 50 Hz, the effect of the surface irregularities such as those shown in Figs 5.7 and 5.8 would be non-existent. So, in acoustical terms, a surface which can be highly non-uniform at 1600 Hz can be seen to be absolutely regular at 50 Hz.

5.6 Low frequency considerations

As discussed earlier in reference to the large room at Blackwing (Fig. 5.2) all 24 m^2 of the ceiling was used as a low frequency absorber. Yet, even with this amount of absorption, the low frequency reverberation time was still much greater than that of the other rooms mentioned. Without such an absorber, the Blackwing room would have produced a build-up of low frequency energy which would have muddied all the recordings, and much definition would have been lost. No such absorption was needed in the Regiestudio room (Fig. 5.5), as such low frequency build-up cannot develop in rooms of such small dimensions because the modal pathlengths are too short to support long wavelength resonances. We shall consider this point further in the following section.

Figure 5.8 shows the way in which the reverberant characters of rooms are controlled by the different acoustical properties of their dimensions and surfaces. At the high frequencies, the room response is generally controlled by the relationship of specular reflexions to absorption. Effectively, here, the sound can be considered to travel in rays, like beams of light. In the mid-frequency band, control is mainly down to the diffusion and diffraction created by the irregularities and edges of the surfaces. At the upper-low and lower-mid frequencies, the room response is mainly that which can be reinforced by modal energy. This region is usually dealt with in terms of normal wave acoustics. As the room size reduces, it tends to produce a greater spacing of modal frequencies, and hence the energy is concentrated in more clearly defined frequency bands which leads to a more 'coloured' or resonant sound characteristic. However, as the room dimensions are reduced, the lowest modal frequencies which can be supported are driven upwards, so excessive low frequency build-up becomes less likely.

Small rooms tend to sound 'boxy' because the more readily defined modal energy is in a higher frequency range, reminiscent of the sound character of a large box, hence the 'boxy' sound. The modally controlled frequency region is bounded in its upper range by a limit known as the large-room frequency[1], and at its lower range by a region known as the pressure zone. The pressure zone can be thought of as a frequency region in which the room is too small to be able to support modal resonances, and is discussed further in Chapter 7. In Fig. 5.9(a), a sound wave can be represented by a line crossing the room. This is a snapshot in time, and shows the positions of high and low pressure for that instant. Another snapshot taken a few moments later would show the peaks and troughs uniformly shifted in the direction of travel. At resonance, however, the interference pattern of the direct and reflected waves becomes fixed in space. A resonance occurs when the distance between two opposing, reflective surfaces bears an exact relationship to the wavelength of the resonant frequency. The modal pathway tends to trap the energy, with the peaks and troughs of the direct and reflected waves exactly coinciding with each other. The energy build-up in such circumstances can be very great. The effect is analogous to a child on a swing. If the energy input to the swing (a push) is timed to coincide with the peak of its travel, then the swing can oscillate strongly to and fro with what appear to be only minimal inputs of applied force. (See Fig. 5.9(b).)

Below these frequencies however, it can be seen from Fig. 5.9(c) that when wavelengths become sufficiently long, there are no noticeable alternate regions of the room with above and below average pressure, but rather the whole volume of the room is simultaneously either rising or falling in pressure. The room response is thus unable to boost the initial sound wave by adding any resonant energy, because it cannot provide any resonant pathways, and the sound thus decays rapidly. This is the reason why the low frequency reverberation response of rooms falls rapidly as room size reduces. Conversely, in large halls, where the pressure zone frequency is in the infrasonic region, even the lowest audible frequencies can be reinforced by modal energy. These frequencies are notoriously difficult to absorb, and are not significantly lost through doors or windows unless these are proportional to their wavelengths. They also refuse to be much thrown from their paths by obstructions, unless these obstructions are of sizes in the region of a quarter of a wavelength of the frequencies in question. In smaller rooms, however, with the rise in the pressure zone frequency, added to the increased losses due to the greater number of wall strikes per unit of time, it is therefore little wonder that the low frequency reverberation time tends to fall off more rapidly.

Figure 5.9 The pressure zone

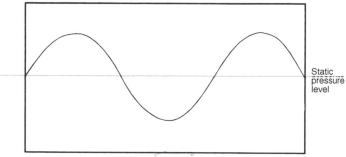

Distance travelled

(a) A wave traversing a room will produce areas of high and low pressure which correspond to the compression and rarefaction half-cycles of the wave. Upon striking the opposing wall, it will reflect back, and if the pathlength is an integral multiple of the wavelength, the returning reflexions will be exactly in phase with the drive signal

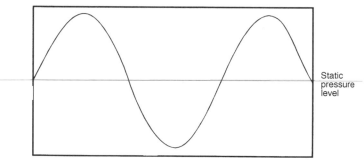

Distance travelled

(b) As the drive signal continues, it will add in-phase energy to the resonant energy, which will build up the total energy in the resonant mode. This is similar to the effect of adding energy to (pushing) a child on a swing at the right moment, when the arc of swing can be made to increase

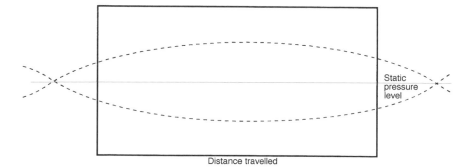

Distance travelled

(c) Where the wavelength is longer than twice the size of the room, the whole room will be rising or falling in pressure, more or less simultaneously. This is the pressure zone condition, where alternate regions of high and low pressure are not evident

5.7 Summary

The rooms depicted in Figs 5.2 to 5.6 all have very different sonic characteristics which are derived from their physical sizes, shapes, and the nature
of the arrangement of their surface materials. This is despite the fact that,
except for their floor surfaces, they are basically all constructed from the
same materials. In these cases, the floor material was of relatively little
overall significance, although stone floors, perhaps, would add just a little
extra brightness.

Rooms such as the ones described here, as with all live rooms, take some
time to get to know, but once known they can be very acoustically productive. They also take a great deal of experience to design and construct. Bad
ones can be really useless, or of such restricted use that they are more of a
liability than an asset. There is no current hope of computer modelling these
rooms, partly because the complexities of the interactions are enormous, and
secondly because the influence of complex room shapes on the acoustical
virtues of the subjective perception are not yet sufficiently understood to programme them into a computer. The influence of the great variety of equipment and people who may 'invade' the acoustic space is also beyond current
modelling capability. However, for engineers wishing to record in such
rooms, hopefully this chapter will have provided enough insight to make the
process more productive.

Note

1 The large-room frequency, bounding the frequency region dominated by normal modal
energy, and an upper frequency range which is mainly diffusion/diffraction controlled, can
be estimated from the simple equation given by Schroeder:

$$f_L = \frac{K\sqrt{RT_{60}}}{V}$$

where: f_L = large-room frequency (Hz)
K = constant: 2000 (SI)
V = room volume (m^3)
RT_{60} = apparent reverberation time for 60 dB decay (seconds)

Chapter 6

Orchestral rooms

Orchestral music was designed to be performed live. When many of the great classical works were written there was no such thing as recording, so the instrumentation and structure of the music was aimed only at its performance in spaces with audiences. Transferring the performance into a studio, perhaps of only just sufficient size to fit the whole orchestra, imposes a completely different set of conditions. As was discussed in Chapter 4, a big constraint on the achievement of a natural orchestral sound in studios is the fact that the whole process is usually entombed in a massive acoustic containment shell. This removes even further the ambience of the studios from that of the concert halls.

Today, much orchestral recording is for the soundtracks of films, and in such circumstances, when the conductors need to see the films as well as to be in close contact with the musicians, then the facilities of a studio are more or less mandatory. On the other hand, under less technically demanding circumstances, ever since the earliest days of recording, a large proportion of orchestral recordings have been made outside of purpose-built studios. However, when one begins to consider the deeper element of achieving good orchestral recordings, the above fact comes as no surprise.

6.1 Choice of venues and musicians' needs

Around the world there are some very famous and widely used locations for orchestral recording. Not too surprisingly, concert halls are one member of this group, but assembly halls, churches and cathedrals are also popular locations. One requirement of such a space is that it needs to be big enough to house the orchestra, but usually, the apparent *acoustic* space needs to be even larger.

If we first consider the obvious, recording in a concert hall, there are two conditions likely to be encountered; recording *with* or recording *without* an audience. When an orchestral recording is being performed for recording purposes, in almost all cases, the conductor will want to discuss the music and his interpretation of its performance with the musicians. It may be necessary to run through separate sections until the right feel is achieved, and errors or misinterpretations in the playing may be pointed out and corrected. Performances intended for recording tend to be auditioned in greater detail

than live performances, and a small detail which may pass in a live performance may become irritating on repeated listening to a recording. The
straightening out of these small points can be very time consuming. Clearly,
much of this is private and personal, and it is usually all conducted in an
atmosphere of professionalism, trust and understanding. Frank and open discussion of these points would rarely be possible in front of an audience, nor
would it be likely to constitute an interesting spectacle for the paying public,
so almost out of necessity, many such recordings take place out of public
view. Unfortunately, if such recordings are made in a concert hall, then that
hall will probably have been designed to have an appropriate reverberation
time when full. Its empty acoustic may not have been a prime object of its
design, although, in an attempt to make the acoustics independent of audience numbers, some halls have empty seats with absorption coefficients
which try to match that of an occupied seat.

With recording techniques using close microphones this may or may not be
of great importance, but for recordings with more distant microphones, the
ambience of some empty halls may be undesirable. However, surprising
though it may seem, even when full, not all concert halls are highly rated for
orchestral recording purposes. A concert hall is an expensive thing to build,
and few can be dedicated solely to orchestral performances. Consequently,
ideal orchestral acoustics may have to be compromised to accommodate other
uses of the halls, such as for conferences, operas, electrified music concerts or
jazz performances. Each use has its optimum set of acoustic conditions, both
in terms of hall acoustics *and* stage acoustics. Furthermore, in addition to the
optimum reverberation requirements, lateral reflexion characteristics may
also form another subdivision of the ideal acoustics for each use.

Some of the old halls, built before recording existed, are still very well
liked. In fact, this is not too surprising, for they frequently did not have to
make quite so many performance compromises as halls of more recent construction. Moreover, when much of the classical music repertoire was
written, it was written with many such concert halls in mind. The fact that
much classical music can be expected to sound good in those halls is something of a self-fulfilling prophesy. Live performances were the *only* performances to be heard by the public when those halls were built, and no
compromises needed to be made to allow for electric amplification or for
other uses. Many orchestral pieces were even written with specific concert
halls already considered for their first public performances. However, this is
somewhat akin to writing and recording a piece of music in a studio with
idiosyncratic monitoring: it may not easily transfer to more acoustically
neutral surroundings.

The great composers usually also had a comprehensive understanding of
the needs of the musicians. Orchestral musicians need to hear themselves in
a way that is both clear and sonorously inspiring. They also need to hear
clearly many of the other musicians in order to develop the feel of the performance. Composers often bore these facts in mind when arranging the
instrumentation, so the music, the instruments themselves and the halls in
which they frequently performed developed not in isolation but in concert
with each other. It is thus little wonder that many of the shoe-box shaped
halls, which have been commonplace for centuries, are still well liked by
musicians, recording personnel, and audiences alike.

In recent years it has become apparent that, for a rich sense of spaciousness, lateral reflexions are of great importance. The shoe-box shaped designs provide plenty of these, but they can be very problematical for many other purposes for which the hall may be used, often playing havoc, for example, with the intelligibility of the spoken word. Conferences or speeches in such halls can be nightmares. In fact, the reason why so many religious masses are chanted and not spoken is because the longer sounds of a chant are less easily confused by the reflexions and reverberations of most churches than would be the more impulsive consonants of normal speech. Hard constants would tend to excite more modal resonances, as they contain more frequencies than the softer chanted consonants. Ecclesiastical chanting therefore appears to have been of acoustic origins rather than religious ones.

Nowadays, however, we must perform much of the old music in multifunctional halls, or in churches, town halls, and the like. What all of these locations have in common though, which sets them apart from most recording studios, is the large amount of space which they have for the audience or congregation. They also tend to have many windows and doors, which allow an acoustic coupling of the internal spaces to the outside world. They are *not* constrained within the bunker structure of a sound isolation shell, and all, therefore, have room to 'breathe' at low frequencies; even if this is not immediately visually apparent from the interior dimensions of the rooms themselves. Given this acoustic coupling through the structure, they tend to appear to be acoustically larger than they are physically.

6.2 RT considerations

The optimum reverberation time (RT) for orchestral recordings tends to vary, according to the music and instrumentation, from about 1.8 to 3 seconds in the mid-band, with the 5 kHz RT usually being around 1.5 seconds. The low frequency RT is a continuing controversy, with differences of opinion as to whether the 100 Hz RT should rise or not, and if so, by how much. In general, there are compromises between definition, warmth and majesty. As the RT rises definition is lost to warmth, which in turn changes it to a character of majesty, until at higher RTs, all is lost to chaos. Inevitably the optimum requirements for each piece of music or each orchestral arrangement will call for somewhat different conditions, as will the presence or absence of an organ, chorus, solo piano, or number of contrabasses in the orchestra. It has certainly been my experience that a moderate rise in RT at low frequencies is usually desirable; but that introduces yet another variable – personal preference.

It was pointed out to me some time ago by the New York acoustician Francis Daniel that the Fletcher-Munson curves of equal loudness could have something to do with the LF debate. Figures 6.1 and 6.2 show the classic Fletcher-Munson curves, alongside the more modern equivalent, the Robinson-Dadson curves. The latter are now generally accepted as being more accurate; however, the differences are too small to have eclipsed the former in general usage. What they both show is how the sensitivity of the ear falls at the extremes of the frequency range. If the 0 dB threshold line at 3 kHz is followed to 30 Hz, it strikes the sound pressure line at 60 dB. If the line passing through the 25 dB point at 3 kHz is followed down to 30 Hz, it

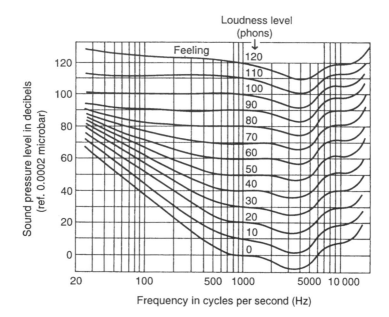

Figure 6.1 The classic Fletcher and Munson contours of equal loudness for pure tones, clearly showing higher levels being required at high and low frequencies for equal loudness as the SPL falls. In other words, at 110 dB SPL, 100 Hz, 1 kHz and 10 kHz would all be perceived as roughly equal in loudness. At 60 dB, however, 10 kHz and 100 Hz would require a 10 dB boost in order to be perceived as equally loud to the 1 kHz tone

will be seen to pass through the SPL of around 65 dB. These are curves of equal loudness, and the above observations mean two things. Firstly, that 60 dB more (or one million times the acoustic power) is needed at 30 Hz to reach the threshold of audibility than is needed at 3 kHz. An extra 40 dB (or 10,000 times the power) is needed at 30 Hz to sound as loud as a tone of 25 dB SPL at 3 kHz. So it can be seen that the ear is vastly more sensitive to mid-frequencies than to low frequencies at low sound pressure levels. Secondly, the rise of 25 dB in loudness from 0 dB SPL to 25 dB SPL at 3 kHz needs only a 5 dB increase at 30 Hz to produce the same subjective loudness increase.

Looking at these figures again, 25 dB above the threshold of hearing at 3 kHz will sound as loud as 5 dB above the threshold of hearing at 30 Hz. The perceived dynamics are considerably expanded at the low frequencies. At high SPLs above 100 dB, the responses are much more linear. Now let us consider a symphony orchestra in a hall, playing at 100 dB. The direct sound will, according to the 100 dB lines on the equal loudness curves, be perceived in a reasonably even frequency balance. However, when the reverberant tail has reduced by 50 dB, and is still very clearly audible at 50 dB SPL in the mid-range, the lower octaves will have fallen below the threshold of audibility.

Considering the fact that much of the reverberation that we perceive in concert halls is that of the tails, after the effect of the direct and reflected sounds has ceased, then much of our perception of that reverberation will be

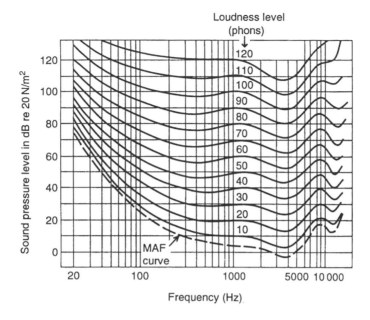

Loudness level
(phons)

Figure 6.2 Robinson-Dadson equal loudness contours. These plots were intended to super-
sede the Fletcher-Munson contours, but, as can be seen, the differences are too small to
change the general concept. Indeed other sets of contours have subsequently been published,
as further updates, but for general acoustical purposes, as opposed to critical uses in digital
data compression and noise-shaping, the contours of Fig. 6.1 and the above suffice. The
lower MAF (minimum audible field) curve replaces the '0 phons' curve of the older,
Fletcher-Munson curves. The MAF curve is not absolute, but is statistically derived from
many tests. The absolute threshold of hearing varies not only from person to person, but with
other factors such as whether listening monaurally or binaurally, whether in free-field conditions or in a reverberant space, and the relative direction of the source from the listener. It is
therefore difficult to fix an absolutely defined 0 dB curve

in the area where the low frequencies would be passing below audibility if a
linear reverberation time/frequency response existed. This suggests that halls
with a rising low frequency reverberation time may be preferable for music
which is performed at lower levels, in order to maintain a more evenly *perceived* frequency balance in the reverberation tails. Rock music on the other
hand, played at 120 dB in a hall, may well degenerate into an incomprehensible confusion of low frequency wash in halls in which the RT rises at low
frequencies. In fact, electrified rock music usually requires a much lower
overall RT than orchestral music, and, quite definitely for rock music, a more
linear RT/frequency 'curve' is desirable, or even one where the low frequency RT falls.

6.3 Fixed studio environments

As has probably already been gathered from all of this, to produce a
large single recording studio to mimic all of these possibilities would be a
mighty task indeed. It is one very significant reason why so many orchestral

recordings are done in the aforementioned wide range of out-of-studio locations.

Live performances inevitably must be recorded in public halls, though the reasons for recording live are not always solely for the hall acoustics. It can be very difficult to lift the definitive performances out of orchestras in studios. The adrenaline of a live performance will, in most cases, give the recording an 'edge' which the studio recordings frequently fail to achieve. Nevertheless it is still heard from musicians that far too many studios also fail to inspire a performance from an ambient point of view. All too often, the 'technical' environment of many studios is not conducive to the 'artistic' feel of a hall. Comfort and familiarity with the surroundings can be a very important influence on musicians, as can their own perception of their sound(s). In most concert halls, or assembly halls and such like used for recording, the performing area is usually surrounded on at least three sides by reflective surfaces. As mentioned before, these lateral reflexions are of great importance to the sense of space, and by association, with the sense of occasion. If the acoustic conditions in a studio can help to evoke the sense of the occasion of a concert stage, then it may well be off to a good start in terms of the comfort factor for musicians.

The one big obstacle in re-creating realistic performance-space acoustics in a studio is the normal lack of the open space in front of the orchestra. During concerts, the musicians face from the platform into a hall from which there will probably be no significant reflexions, but much reverberation. The musicians thus receive the bulk of their reflexions from the stage (platform) area, and the subsequent reverberation from the hall in front of them. Human hearing is very directional in the horizontal plane, so the perception of the necessary reflexions, which help to reinforce and localise the instruments, can be easily discriminated from the equally necessary reverberation. The reverberation does not swamp the reflexions because of the spacial separation of the two.

In a studio of practical size, we have a great problem in trying to re-create this situation. If the orchestra face a reflective wall, the reverberation will be confined within the area of the performance, and additional reflexions will also arrive from the front wall. If we make the wall absorbent, we will tend to kill the overall reverberation. Either result is unnatural for the orchestra. It seems to me that perhaps the only way to overcome this would be with a relatively absorbent wall containing an array of loudspeakers, coupled to a programmable artificial reverberation system. This way a reasonably sized studio could be created, say about the size of a concert platform, but with the effect of a reverberant hall in front of the musicians. The directional characteristics of the sound could thus be preserved whilst keeping the studio to a reasonable size.

It is doubtless the directional sensitivity of our hearing which is at the root of the phenomenon that when using conventional microphones (SoundField microphones *can* be an exception to this) the necessary microphone position for a subjectively most similar direct/reflective/reverberant balance is usually considerably closer to the orchestra than would be the corresponding seating position for a listener. However, it must be said that the microphone positions are usually above the audience, so they perhaps suffer less from local absorption, but this does not account for all of the effect.

What tends to set orchestral studios apart from most other studios is that a generally rectangular acoustic shape is rather more common. Walls are normally faced with a significant amount of diffusive objects to break up unwanted cross-modes, but the more or less parallel-sided nature of the walls can help greatly in the reproduction of the environment of the platform of a hall. Ceilings, though, as in halls, are almost never parallel to the floors. In some multi-functional rooms, they *may* be found to be parallel, but they are most likely to be relatively absorbent. Many concert halls have acoustic reflectors above the platform, usually angled to project more of the sound towards the audience, though frequently the sound also rises into scenery and curtain raising mechanisms, which can be absorbent in their overall effect.

However, in no cases can studio acoustics be separated from the polemic on the subject of the use of close versus distant microphones. I personally have many dozens of classical CDs, and must admit in many cases to enjoying the dynamism of the close mic'd recordings when I am listening to them, yet I can equally enjoy the more integrated grandeur when I listen to many of the 'stereo pair', distant mic'd, or SoundField recordings. The problem is, of course, that a listener at a live concert cannot be both near and far from the orchestra at the same time. With microphones, via a recording, it *is* possible, but to my ears the result is perceptually confusing unless either one is greatly subservient to the other, and present only for added detail or richness. The problems of which option to take usually revolves around so many of the fundamental problems of psychoacoustic perceptions, some of which are mutually exclusive. Essentially, close and distant microphone techniques are two entirely different things, and the use of either is a matter of choice. They are not in competition for supremacy, but merely options for the producers, who will choose the technique which is considered to be appropriate for any given occasion.

6.4 Psychoacoustic considerations and spacial awareness

Not all of the problems of the design of recording studios for orchestral use are of direct concern to the recording engineers. Many other things which affect only the comfort and sense of ease of the musicians must be taken into account if the best overall performances are to be recorded. Sometimes these things can be better explained by looking at extreme situations, which help to separate any individual characteristics which may only be subconsciously or subliminally perceived in the general confusion of the recording spaces. So let us now take a look at some of the spacial effects which can play such great parts in the fields of awareness and comfort.

I remember my first visit to the Anglican Cathedral in Liverpool, UK. I went to hear its pipe organ, which was originally commissioned in 1912 as the largest and most complete organ in the world. It has 145 speaking stops, over 30 couplers, around 9700 pipes, and on Full Organ can produce a 120 dB SPL in the 9 second reverberation time of the cathedral. It is blown by almost 50 horsepower of blower, and it needs it all, because the cathedral is one of the ten largest enclosed spaces in the world. Its size is truly awesome, and a well-known producer accompanying me at the time admitted to being frightened by its overwhelming size.

When inside the cathedral, one is aware of a reverberant confusion from a multitude of noise sources, but occasionally, on windless days, and when not open to the public, there are occasions of eerie silence. When speaking in a low-to-normal voice to a person close by in the centre of the great nave, it is almost as though one is speaking outside in a quiet car park. It is not anechoic, as there is a floor reflexion to liven things up, but the distances to the other reflective surfaces are so great that by the time they have returned to the speaker, they are below the threshold of audibility. On the other hand, a sharp tap on the floor with the heel of one's shoe produces a sharp rap, followed almost half a second later by an explosion of reverberation. The clarity of the initial sound is absolute, as coming from the floor, it has no floor reflexion component; yet a subsequent tap, originating before the reverberation of the first one has died down, is almost lost in the ambient sound. This highlights well the temporal separation discrimination which plays so much part in orchestral acoustics.

Back in 1982, at a time when I was still involved in the recordings of Mike Oldfield, I was visited by Hugo Zuccarelli, the Argentinian-Italian 'Holophonics' inventor. He came to a studio where I was recording to see if I thought that Mike would be interested in using 'Holophonics' on his recordings. Mike never used it, but Pink Floyd and others certainly did. On headphones at least, Hugo could pan things around the head with stunning realism; and a realism not only of position but of clarity as well. He also had a demonstration cassette of rather poor quality, yet the sounds of jangling keys remained crystal clear in front or behind the head, whilst the tape hiss remained fixed between the ears. It was an impression that I will never forget, as the wanted sounds had become absolutely separated from the noise, existing totally natural and noise-free in their own spaces. This demonstrated most clearly to me the powerful ability of the ear to discriminate spacially, even in what amounts to an entirely artificial environment. The holophonics signal supplied strong phase-/time-related signals to the ears, not present in mono signals, nor necessarily in multiple microphone stereo recordings. The extra phase/time information provides a powerful means for the brain to discriminate between different signals, including between wanted and unwanted sounds.

The old chestnut of the 'cocktail party effect', to which the holophonics effects closely relate, has been around for years, and is no doubt already familiar to most readers. To recap, if a stereo pair of microphones is placed above a cocktail party, and auditioned on a pair of headphones, but panned into mono, then the general murmur of the party would be heard, but it would be difficult to concentrate on any individual conversation unless one of them happened to be occurring very close to the microphones. However, when auditioned in stereo, at exactly the same level, suddenly it becomes clear that many separate conversations can be easily recognised and understood. The precise audiological/psychoacoustical mechanisms behind this have still not been fully explained, but nonetheless the effect itself is very well established as fact.

If we lower the level of the cocktail party recording into the domain of the background noise of the tape hiss, then the conversations will begin to get lost in it, masked by the hiss even whilst still relatively well above the absolute hiss level. But such is not the case with the effect of holophonics, which can lift a recognisable pattern out of the hiss by spacial differentiation. The phenomenon, once recognised, has a lasting effect. The significant difference between the effects of holophonics and conventional stereo is that conventional stereo only provides information in a single plane; the holophonics effect is three-dimensional. The single plane of conventional stereo contains all the information, both wanted and unwanted, so the tape hiss shares the same sound stage as the wanted signals: they are all superimposed, spacially. In holophonics, however, the wanted signals can be positioned three-dimensionally, yet the tape hiss is still restricted to its single plane distribution. Once sounds are positioned away from the hiss plane, they exist with a clarity which is quite startling, and it is remarkable how the hiss can be ignored, even at relatively high levels, once it is *spacially* separated from the desired signals.

We thus have several distinct mechanisms working in our perception of the spaces that we are in, and all of the available mechanisms are available to the musicians in their perceptions of the spaces in which they are performing. Clearly, all of these mechanisms cannot be detected by microphones, nor can they be conveyed to the recording medium, and consequently, many of them will never be heard by the recording engineers in the control rooms. In fact, although the recording engineers may well, consciously *or* subconsciously, perceive these things, they often fail to fully realize their importance to the musicians. To the musicians, however, who work with and live off these things daily, their importance cannot be overstressed. In the recording room therefore, all of these aspects of the traditional performance spaces need to be considered. At the risk of becoming repetitive, I say again, if the musicians are not at ease with their environment, then an inspired performance cannot be expected.

From time to time, there are situations which require even orchestral musicians to wear headphones when recording; either full stereo headphones, or single headphones on one ear. Perhaps the major reason why this is usually avoided is not the enormous number of headphones required, but that inside headphones, certainly if they are of the closed type, it is not possible to reproduce the type of complex sound-field that the orchestral musicians are accustomed to playing within. Human aural pattern recognition ability is very strong, often even when sounds are hidden deep within other sounds, and even deep in noise. If headphones isolate the musicians from the usual patterns, the effects can be very off-putting.

For many of the reasons discussed in this chapter, orchestral studios should not be designed primarily simply for what is good from the point of view of the recording personnel. First and foremost, a performing environment should be created to aid the *whole* recording process. I realise that there are hi-fi fanatics who disagree, but for me, I would much rather hear a compromised recording of an excellent performance, than an excellent recording of a compromised performance. Obviously, though, the real aim of the exercise is to make excellent recordings of excellent performances, and this is why so many things must be considered.

I must admit that I did not fully appreciate the importance of the recording environment, or the extreme fussiness of some musicians about their foldback balance, until, when producing a recording, I was having problems getting some backing vocalists to sing with the appropriate phrasing. I was still unhappy with the performance after they had left, and was singing along with the tape, showing the engineer what I meant, and seeing if he agreed. He not only agreed, but said that my voice also suited the part, then encouraged me to go into the studio to try to see how it fitted the track. In the space of minutes, foldback had suddenly, for me, become the most important thing in the whole process. If the overall sound was too loud, I could not phrase properly, if it was too quiet, it was difficult not to swamp it by my own voice inside my head. If my voice was too low in the track, I strained, and tended to sing sharp; and if it was too high in the track, I held back, lost my dynamics, and had a tendency to sing flat. From that day onwards, I suddenly saw foldback in a very different way. The experience was quite a shock, and I felt a terrible guilt about the many times in the past when I had perhaps been a little less considerate to the musicians on that subject than I should have been.

Many musicians of all kinds play off their own sound; it is like a feedback mechanism that both reassures and inspires them. This is true of more or less every musician, and of whatever type of music that they perform. If their sounds are not given to them in their foldback as they need, then it can be very disconcerting. In the case of orchestral musicians however, their foldback is not usually via headphones, but via the reflective surfaces of the room. If playing to a backing track, it will possibly only be the conductor who will listen to it on headphones. In these cases, the acoustics of the performing space provides the foldback to the musicians, so it should be given its due consideration as such.

6.5 Use of screens

Another frequent conflict between the needs of musicians and those of the recording engineers is the use of acoustic screens. In the close-microphone type of recording, separation is a factor which is often considered desirable by the engineers. This can be improved by the use of acoustic screens between different sections of instruments, which normally, as a concession to the musicians, have windows to allow eye to eye contact. Unfortunately, this can disrupt the perception of the desired acoustic sound-field by the musicians, and in most cases, they would prefer the screens not to be there.

It has been quite amazing how, over the course of recording history, musicians' needs have often been neglected in the recording process. Time and time again, if the recording engineer has had a problem with overspill, screens have been imposed upon the musicians, without due consideration of the effects of their insertion on other aspects of the recording process. In many cases, recording staff have totally failed to appreciate the artistic damage which can be done by delays and disturbances caused by technical adjustments. Awareness of such things has been one thing which has set apart the specialist recording engineers and producers, who by understanding things in a more holistic way, have gained the co-operation and respect of the musicians, and have thus produced more inspired recordings from which they have deservedly built their reputations.

6.6 Summary

This discussion on some rather peripheral aspects of audiology and psychoacoustics has not been a digression, but is a fundamental requirement in understanding what is necessary for the design of good recording spaces for orchestral performances (or perhaps rather, for the design of good *performance* spaces for orchestral recordings). What must have become evident from the discussion is that acoustic variability is a prime consideration in any such design, unless the space is to be restricted to excelling in the recording of perhaps only a small range of the likely performances for which it may be required.

The room variability needs to be in terms of overall reverberation time, and if possible, the relative balance of low and mid/high frequency reverberation. Reflexions need to be controllable in terms of time, direction, and density, with the availability of some reasonably diffusive surfaces to add richness without undue colour. Always, however, the consideration of what the musicians need should be given at least equal weight to the needs of the recording staff when acoustic adjustments are being made. It is very necessary to strike a balance between these priorities, and to achieve the close co-operation between all parties involved. Always remember, the variable acoustics are not just provided for the benefit of the recording staff.

Chapter 7

Vocal rooms

7.1 Aims

When listening to recorded music, and especially on headphones, one frequently becomes aware of the sound character of the room in which the vocals were recorded. In itself, this should not be a problem, unless the room sound is 'boxy' or inappropriate to the song or the rest of the instrumentation. Unfortunately though, this is all too often the case, as the vocals have either been recorded in a small 'vocal booth' to achieve the desired separation from the other instruments, or they were recorded in a small room, perhaps because of convenience. The reason could also possibly have been because no large or neutral rooms were available. What is also unfortunate is that far too many control rooms, and/or monitor systems, are insufficiently neutral in themselves to allow the recording staff to notice the subtleties of the vocal room acoustics. This is especially the case in many multimedia or 'project' studios, where due attention to control room monitoring conditions has often been sadly lacking. Vocal room acoustics often have decay times which are less than those of most control rooms, hence the sound of the recorded room becomes lost in the monitoring acoustics of the control room. In other cases, the staff of a studio can become so used to the sound of a vocal room, that they no longer hear it in the background of their recordings. The problem is perhaps now greater than it used to be, as the lower noise floors of digital recording systems have rendered audible, in the home, sounds which would in earlier years have been lost in the background noise.

Where neutrality is required, vocals are usually best performed in the middle of large rooms, where there are few early reflexions. If floor reflexions are a problem, then rugs can always be provided for the vocalists to stand on, and this also helps to reduce the pick-up of the sound of any foot movements. However, most of the vocal energy usually tends not to be directed towards the floor, and so very little returns to cause problems. What is more, the selected microphone patterns are usually either cardioid or figure-of-eight, and hence naturally tend to ignore the floor reflexions.

Purpose-designed vocal rooms usually need to be as neutral as possible, unless the acoustic liveliness of a room is being used for effect. The problem with using *live* rooms for vocals is that, usually, the nature of the room ambience that is considered good for instruments is not an ambience that does much for voices. In large rooms, the space around the microphone is usually conspicuous by the absence of any early reflexions which can colour the sound, but in small rooms, this is not so easily achieved. As general neutrality is difficult to achieve in small rooms, and especially so in very small rooms of the size commonly associated with vocal rooms, then in circumstances where no large room is available it is perhaps better for vocal recordings to be made without any room ambience whatsoever, except perhaps for the reflexions from the floor and a window. With the use of cardioid or figure-of-eight microphones, they can usually be positioned so as to minimise the pick-up of any of these floor or window reflexions. A further difference between the recording of vocals, and the recording of many other instruments, is that there is an intelligibility factor to be considered. Many 'vocal gymnastics' would be rendered unintelligible in rooms where insufficient 'space' existed around the vocal sound before the reflexions returned to the microphone.

Due to the general persistence of the energy in low frequency room modes, simple attempts at absorption by the placing of acoustically absorbent tiles on the walls and ceiling will not suffice. These will tend to absorb the higher frequencies, but leave the lower-mid and low frequency modes largely untouched. This will produce a room with a heavily coloured ambience, which will lack life and add a thickness to the sound, robbing it of much clarity. Rolling out the offending frequencies by equalisation will take the lower frequencies out of the unwanted ambience, but it will also take them from the direct sound of the voice. In turn, this will disturb the natural harmonic structure, and will tend to remove much of its power and body. As discussed in Chapter 2, to make a very small, musically neutral room is virtually impossible, so, in the vast majority of cases, the only thing to do with a small vocal room is to absorb everything, and then provide a few discrete reflexions.

If we make the room too dead, the vocalists, on entering the room, may find it uncomfortable. In almost all cases, the vocalists will wear headphones when recording, but nonetheless their initial impressions when entering the room can have a lasting impression. They should never be allowed to feel uncomfortable, even if only for the few seconds between entering the room and putting on the headphones, as it is remarkable just how these few seconds can leave the vocalists unsure of their vocal sounds. Illogical it may well be, but artistic performances tend to be fragile things, and anything which runs the risk of introducing any extra insecurity into musicians is to be avoided. Fortunately, a hard floor and a window, or glass door, can provide sufficient reflective life to avoid the vocalists experiencing an anechoic chamber effect when they enter the room, and this is usually achievable without creating a boxy character.

7.2 Practical constructions

Figure 7.1 shows a layout for a vocal room which occupies only about 9 m² of floor area and 3 m of height. It assumes a 'worst case' structural shell of what is more or less a 3 m cube. We will presume that one wall adjoins a control room, and that the other surfaces are concrete. In many cases, vocal rooms are not only used for singing, but can also be used for voice-overs or dialogue replacement. In these cases, there would be no music to mask any

Figure 7.1 Vocal room

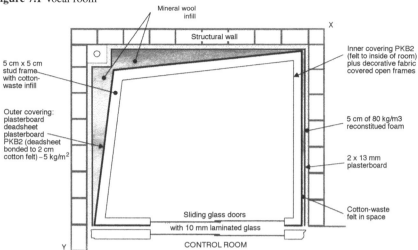

(a) Plan

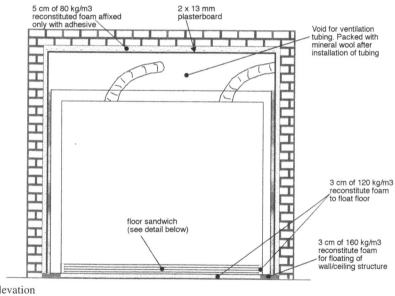

(b) Elevation

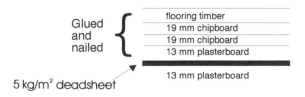

(c) Detail of floor sandwich, laid on top of reconstituted foam

extraneous noises, so the room should be well isolated. In the case of Fig. 7.1, which is based on the design of a very successful room, the whole structural shell was first lined with a 6 cm layer of reconstituted polyurethane foam of 80 kg/m^3 density, except for the floor, where a 3 cm layer of a 120 kg/m^3 variety is used.

The reason for the higher-density foam for the floor is because of its greater resistance to flexure under load. A thicker, lower-density foam would provide equal isolation, but if a heavy amplifier were to be placed in one corner, the floor would be less stable than if the construction method shown in Fig. 7.1 was used. Figure 7.2 shows the effect of uneven floor loading when using lower density foam with this type of separately floated floor. Figure 7.3 shows an alternative construction method, where the walls are mounted *on* the floated floor. In this case, if a greater thickness of low-density foam is used, the weight of the walls and ceiling will bear down on the edges of the floor, locally compressing the foam and creating a 'crown' in the middle. There are pros and cons to each method of construction. The first method completely decouples the walls from any impact noise directly impinging on the floor, such as if the room is being used for a bass drum, or bass guitar amplifier, when this method yields better sound isolation. The second method imparts higher damping on the whole floor, and thus reduces any resonances that may exist. Incidentally, vocal rooms of the type being described here do tend to be excellent for the recording of bass guitar amplifiers, especially if a sound with tight impact is being sought.

7.2.1 Isolating materials – density vs thickness

As for the relative isolation of thicker, low-density foam or thinner high-density foam, then for any given mass, the former is usually better. In general, barring any extra production processes that need to be paid for, one tends with most materials to pay for mass. Prices per kilogram are usually roughly equal. If we were to use a material such as mineral wool, then at

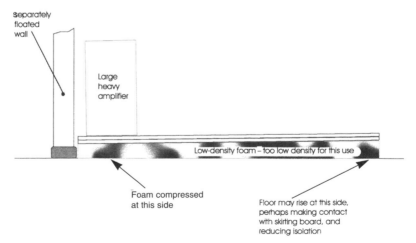

Figure 7.2 Effect of uneven floor loading on low-density float material

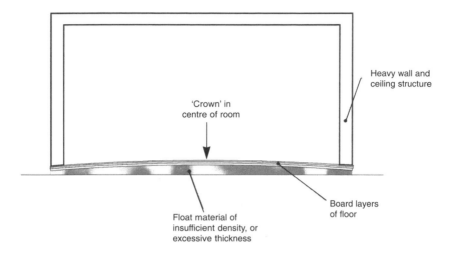

Figure 7.3 Crown effect

125 Hz, a thickness of 1 unit (above a given minimum) could have an absorption coefficient of say 0.07, yet a thickness of four units, for a material of similar density, an absorption coefficient of 0.38, which is over 500% greater. The absorption more or less goes up in direct proportion to the thickness. On the other hand, if we have a given thickness of mineral wool having one density unit, and another having the same thickness but four times the density, our absorption coefficient of 0.07 in the first instance would rise to only about 0.1 in the case of the same thickness but four times the density, which is an increase of less than 50% in effectiveness.

Weight for weight, and hence in rough terms, cost for cost, adding four layers of the lower density material, and allowing it to occupy four times the volume of one layer, would produce over ten times the absorption at 125 Hz than would compressing the four layers into the space of one, i.e. quadrupling the density. Table 7.1 shows some typical figures for different thicknesses and densities of a mineral wool material at different frequencies. Note how at higher mid-frequencies, above a certain minimum thickness, neither increasing the density nor the thickness has much effect.

7.2.2 Speed of sound in gases

In fact, whilst we are dealing with comparative tables, let us look at some of the mechanisms responsible for the effects shown in Table 7.1. Newton originally calculated the speed of sound in air from purely theoretical calculations involving its elasticity and density. He calculated it to be 279 metres per second at 0°C, but later physical experiments showed this figure to be low, and that the true speed was around 332 m/sec. There was conjecture that sound propagation only took time to pass through the spaces between the air particles, and that only the 'solid' particles transmitted the effect instantly. Newton rejected this, but then himself proposed that whilst

Table 7.1 Effect of thickness and density on acoustic absorption

Thickness	Frequency (Hz)			
	125	500	2000	4000
1.25 cm	.02	.12	.66	.62
2.5 cm	.07	.42	.73	.70
5 cm	.19	.79	.82	.72
10 cm	.38	.96	.91	.87

Density 30 kg/m^3

Density	Frequency (Hz)			
	125	500	2000	4000
30 kg/m^3	.07	.42	.73	.70
60 kg/m^3	.09	.60	.75	.74
120 kg/m^3	.10	.70	.77	.76

Thickness 2.5 cm

the sound took a finite time to travel through the particles, the particles themselves did not occupy the entire space in which they existed, and that it was this fact which accounted for the difference. This idea also failed to be convincing. The difference continued to mystify people, until Peire Simon, the Marquis de Laplace, applied what is now known as Laplace's Correction.

It is now common knowledge that air heats when it is compressed, and cools when it is rarefied. As a sound wave travels through air, it does so in a succession of compressions and rarefactions. As mentioned earlier, Newton's calculations were based on the elasticity and density of the air. Elasticity is the ability to resist a bending force and to 'push back' against it, and the speed of sound through a material is partially dependent upon its elasticity. As the compressive portion of a sound wave compresses the air, it increases its elasticity in two ways; firstly by increasing its density, and secondly by the heat which the compression generates. Where Newton had gone wrong was in the omission of the temperature changes from his calculations: he had only taken into account the elasticity increase from the density change. Perhaps this was due to the fact that there is no *average* temperature change as a sound wave passes through a body of air. However, *locally*, there are temperature changes in equal and opposite directions on each compression and rarefaction half-cycle. It is tempting to deduce that the temperature increases and decreases would cancel each other out, which is perhaps exactly what Newton did, but they do not.

As air is compressed, its volume is reduced, and on rarefaction, its volume expands. The internal force which *resists* these changes in volume is its elasticity. If a tube containing air is sealed at one end and fitted with an air-tight plunger towards the other end, then as the plunger is pushed and pulled, the air will compress and expand. When the force on the plunger is released, it will spring back to its resting position. If the tube was then filled with a gas of higher elasticity, the force needing to be applied to the plunger for the same volume changes would be greater, as the higher elasticity would be more able to resist the changes.

It is possible to visualise the transmission of sound through air in a manner similar to that shown in Fig. 7.4. In this figure, a series of ball-bearings are seen contained in a tube, and are separated by coil springs. If the springs are quite weak, then an impact applied to the left-hand side of ball A will pass to B, and on to C, and so forth, but there will be a noticeable delay in the transmission of energy from one ball to the next. The wave will clearly be seen to pass along the tube. Now let us assume that we apply heat to the springs, and the effect of the heat makes them much stiffer. Another impact applied to ball A will again pass down the tube, but with stiffer springs, the wave will travel more rapidly. In fact, at the extreme condition of almost rigid springs, the passage of the impact from ball to ball would be more or less instantaneous, as the ball/spring combination would behave similarly to a solid rod. The speed of the transmission of the force through the system can therefore be seen to be proportional to the stiffness of its springs. Effectively, in air, the particles can be thought of as balls being connected by springs, and that the elasticity of the air is a function of the strength of these 'springs'. The force on one air particle thus compresses the spring, *heating it up*, and hence increasing the elastic force which acts on the next particle. The heating effect caused by the compression thus serves to augment the elasticity of the gas, and hence the speed with which sound will propagate through it.

In the process of rarefaction, if we consider our tube to be of ball-bearings once again, and if the balls are securely attached to the springs, then a rarefaction will pull on ball A, which will in turn pull on ball B, and on to the other balls in turn. At rest, the elastic force on ball B holds it in position, because the springs A–B and B–C are in equilibrium. By pulling ball A away from ball B, the elastic force on B is reduced on the side nearest to A, and thus the greater force from spring B–C will begin to force B towards A, until equilibrium is restored. As B begins to travel towards A, the force B–C will become less, so in turn the extra force D–C will begin to push on C, which will begin to move towards B. The wave of energy will propagate along the tube until all the balls are equally spaced once again, but all shifted slightly further in the direction of the pulling force. This is where, somewhat surprisingly, we find that the cold of rarefaction serves *not* to cancel the heat of compression but to work in concert with it. Hopefully, Fig 7.5 will help to show the concept in a more graphic form.

In rarefaction, the density of the A–B 'spring' is thus reduced, so the force on the C side of B will be greater than on the A side. The cooling produced by the rarefaction will then reduce the elasticity still further (weakening the spring) and thus will act in the *same direction* as the density reduction,

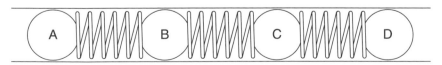

Figure 7.4 Energy transmission via masses and springs. Four balls (A, B, C and D) are separated by springs in a glass tube. The system is shown in equilibrium, with no force applied, and the springs in a relaxed state

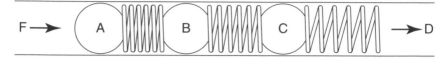

Figure 7.5 Application of force to system shown in Fig. 7.4. If a force is applied to ball A it will transmit the energy to B via the spring. B will then, via the B–C spring, transmit the energy to C, and so forth, until the whole chain of balls and springs is moved in the direction of the applied force. The speed with which the force will pass along the line is proportional to the stiffness of the springs (the elasticity of the coupling). In the above example, the spring between A and B is seen to be compressed. The force on B is thus no longer in equilibrium, so B will move towards C, transferring some of the energy in spring A–B to compress spring B–C until springs A–B and B–C are in equal compression, when B will tend to come to rest. In this state, spring B–C will be partially compressed. C will thus have more compression in the B–C spring than in the C–D spring, so it will move in the direction of D. Due to the applied force and the momentum in the moving balls, the system will oscillate at its natural frequency until it finally comes to rest, when the applied energy has been expended, shifted to the right

making the A–B force even less. This will mean a greater differential between the A–B and B–C forces, so B will be pushed away from C with greater force than that of the rarefaction density change alone.

Hopefully, it can be seen from the above, rather long-winded discussion that the heat of compression and the cold of rarefaction both work in the same direction as the density changes, and hence both *increase* the effect. The additional help caused by the elasticity changes due to heat is the reason for the speed of sound in air being greater than that first calculated by Newton from the elasticity and density alone. The heat of compression and the cold of rarefaction do not cancel each other out but work together to increase the speed of sound through any medium. Once this was realised, experiments were performed to test the ability of air to radiate heat, which was subsequently found to be very poor. This explained why the temperature changes remained in their respective compression and rarefaction cycles, and did not affect each other.

Why we have been discussing all this history following our table of absorption coefficients of mineral wool at various densities and thicknesses is that the above fact is one of the mechanisms at work. In loosely packed fibrous material, such as mineral wool, glass fibre, bonded acetate fibre or Dacron (Dralon) fibre, the fibres can act to conduct the heat away from the compression waves, and release it into the rarefaction waves. This removes a great deal of the heat changes available for the augmentation of the elasticity changes, and hence tends to reduce the energy available for sound propagation. It changes the propagation from adiabatic (the alternate heating and cooling) to isothermal (constant temperature) which slows down the speed of sound. It was the isothermal state that Newton had calculated, but it was the adiabatic state that was the normal situation. Due to the above effects, a loudspeaker cabinet which is optimally filled with fibrous absorbent material appears to be acoustically bigger than it is physically, as the speed of sound within the box is slowed down by the more isothermal nature of the propagation through the absorbent fibres.

Incidentally, Laplace's Correction, as referred to earlier, is quite easy to explain. If we place a known volume of air at 0°C, and at a known pressure, into a pressure vessel, then heat it by 1°C, the air cannot expand, so the pressure will rise. If we then place the same volume of air also at 0°C in a vessel of the same volume, but fit it with a plunger which can be forced outwards as we heat the air to 1°C, the pressure can be allowed to remain constant, by allowing the volume to change. In each case, we heat the same *mass* of air by 1°C, yet the heat necessary to do so in each case is quite different. One is called the specific heat of air at constant volume, and the other, the specific heat of air at constant pressure. In fact, the latter (C_p) divided by the former (C_v) gives an answer of 1.42, and it was the square root of this number by which Laplace multiplied Newton's original calculation in order to correct his calculated speed of sound in air to the observed speed. The reason for the difference between the two specific heats is that in the second instance (C_p) extra heat is consumed in the work of expanding the gas.

7.2.3 Other properties of fibrous materials

Still on the subject of the absorbent properties of our fibrous materials, there are other forces at work besides their ability to convert the sound propagation in air from adiabatic to isothermal. There is a factor known as tortuosity, which describes the obstruction placed in the way of the air particles in forcing them to negotiate their way round the medium. The tortuosity, in increasing the pathlength of the sound which travels through the fibres, also increases the viscous losses which the air encounters as the sound waves try to find their way through the small passageways available for their propagation. In certain conditions, air can be quite a sticky fluid. There are also internal losses as the vibrations of the air cause the fibres to vibrate, and in order to bend, they must consume energy. Frictional losses are also present as vibrating fibres rub against each other. Energy is required for all of this motion of the fibres, and by these means, acoustic energy becomes transformed into heat energy.

Because the above losses are proportional to the speed with which a particle of vibrating air tries to pass through the material, their absorption coefficient is greater when the particle velocity is higher. If we consider a sound wave, arriving at a wall, the wall will stop its progress, and reflect it back. At this point of reversal of direction, the pressure will be great, but the velocity will be zero. The same consideration applies to any ball which bounces from a wall. The absorbent effect of fibrous materials is thus greatest when they are placed some distance away from a wall, and if any one frequency is of special interest, then a distance of a quarter of its wavelength would be an optimal spacing away from a wall for the most effective absorption by a fibrous material. Conversely, membrane absorbers are dependent upon force for their effectiveness, so should be located near to the point of maximum pressure (i.e. close to a wall) if absorption is to be maximised. The absorption mechanisms are thus very different.

7.2.4 Absorption coefficients

Before leaving the subject of fibres and returning to our vocal room, let us consider the concept of sound absorption coefficients, and clear up any doubts about absorption and isolation. Acoustic absorption is the property which a material possesses of allowing sound to enter and *not* be reflected back. In this sense, absorption coefficient refers to sound both internally absorbed, and also that which is allowed to pass through. A large, open window is therefore an excellent absorber, as only minuscule amounts of any sound which reaches it will be reflected back from the impedance change caused by the change in cross-sectional area of the spaces on each side of it. A solid brick wall is a very poor absorber as it tends to reflect back most of the sound energy which strikes it.

Now let us put some practical figures on some different materials. A 2.5 cm slab of a medium-density fibreglass can absorb about 80% of the mid and high frequency sound which strikes it. An open window will absorb in excess of 99% of the sound energy which is allowed to pass through it, and a brick wall, made from 12 cm solid bricks, will allow about 3% of the sound to enter or pass through. Looked at another way, the fibreglass will reflect 20% of the sound energy back into the room, the open window will reflect less than 1%, and the brick wall 97%. If we now look at the same materials in terms of sound *isolation*, the situations are very different. An open window will provide almost no isolation, except a small amount at frequencies with wavelengths longer than the largest dimension of the opening. Our slab of 2.5 cm fibreglass will provide around 3 dB of isolation (although at low frequencies almost nothing), but our brick wall will provide over 40 dB of isolation. Thus in these cases, absorption coefficient and sound isolation properties are unrelated. In fact, in the examples quoted, they run in reverse order. Absorption and isolation properties should not be confused.

7.3 Application to a practical room

Let us look in more detail at the vocal room in Fig. 7.1. As already stated at the beginning of Section 7.2, the walls and ceiling are covered with 6 cm reconstituted polyurethane foam, attached by contact adhesive. In turn, the foam is then lined with a layer (or two if needed) of 13 mm plasterboard. This type of combination, mass/spring/mass (plasterboard/foam/wall), is ideal here as it serves two purposes. Firstly, it provides a good degree of broadband sound *isolation*. Secondly, it also provides a good degree of low frequency sound *absorption*. It therefore acts like a reasonably good combination of our open window and brick wall from the previous paragraph. However, this system achieves by internal absorption what the wall and window do by reflexion and transmission. What is more, if the space between the inner, floated room structure and the foam/plasterboard isolation treatment is lined with a fibrous, high-density material, we can then incorporate the properties of fibrous absorption to prevent any resonances from developing in the gap. We can thus combine high absorption, high isolation (low transmission) and low reflexion, all from the same composite lining. This is important because in our vocal room we do not have sufficient space for large, conventional, wideband absorber systems.

The vocal room 'box' is constructed on the higher-density foam which is first placed on the floor. The walls and ceiling structures are similar in nature to the 'neutralising' acoustic shell described in Chapter 2. This time, however, let us consider the progress of a sound wave as it leaves the mouth of a vocalist, and arrives at the room boundaries as shown in Fig. 7.1.

Remember, we want this room to be sufficiently dead to give no recognisable sound of its own to the recording, but to have just enough life to prevent the room from making the vocalist feel uncomfortable when entering it.

7.3.1 The journey of the sound waves

The sound waves expand from the mouth of the vocalist in a reasonably directional manner. Except at the lowest frequencies, this fact becomes self-evident upon working in these rooms. Another person becomes remarkably quieter if he or she should turn away from you, unless that is, they turn to face the reflective glass doors, or the floor. In most cases, the vocalist *will* be facing glass doors or windows, as visual contact with the control room is usually required for operational purposes. The sound will leave the mouth and strike the glass, which will reflect a good 90% of the energy back into the room with a very wide angle of dispersion. Without headphones, the vocalist will hear this, plus any reflexions from the floor, which may either be direct or via the window. These reflexions will help to alleviate any sense of being an oppresively dead room.

Energy reflecting back from the glass will strike the two side walls and the ceiling with an oblique angle of incidence. Only the wall opposite the glass will receive any sound at perpendicular incidence. The oblique arrival of sound at the face of an absorber will usually cause a much greater loss of energy, as it must pass through the absorbent material in a diagonal manner,

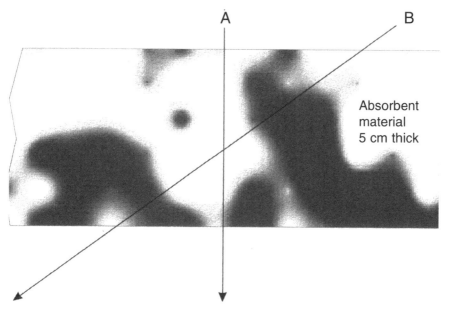

Figure 7.6 Effect of angle of incidence on absorption. An incident wave A, striking the absorbent material at 90° would pass through the 5 cm thickness of the absorbent material. Incident wave B, striking at a shallow angle, would pass through approximately 10 cm of the absorbent material. For material of random texture, the absorbent effect is correspondingly increased in proportion to the extra distance travelled through the material. It should be noted, however, that for certain types of material with a pronounced directionality of fibres or pores, the above case may not apply

and thus effectively travels through a thicker section of material (see Fig. 7.6). Any sound reflecting from the window and the floor will automatically then pass into the walls at oblique angles. If a cardioid microphone is placed between the vocalist and the window, and faces the vocalist, it is only capable of receiving sound directly from the vocalist's mouth, plus, perhaps, a minute amount of the floor reflexions.

The sound which passes into the first layer of felt, behind the decorative fabric surface, will be partially absorbed, but some will continue through to reach the deadsheet backing. At medium and high frequencies, the deadsheet is reasonably reflective, but any high frequencies reflected from it, even at 90° incidences, will still have to pass once again through the felt in order to re-enter the room. For wavelengths in the order of 8 cm (4 kHz), 2 cm of fibrous absorption will be highly effective, as the internal passage of sound through felt at relatively short wavelengths is complex. If only 10% of mid and high frequencies were reflected from the surface of this vocal room on the first contact with a wall, then the second strike would only reflect 10% of that 10%. By the third bounce, which in such a small room may only take 15 or 20 ms, the energy remaining in the reflexions would only be thousandths (10% of 10% of 10%) of that leaving the vocalist's mouth. It would thus be 30 dB down within the first 20 ms, and 60 dB down in considerably less than 50 ms.

At low frequencies, such as those produced if a bass guitar amplifier were to be placed within the room, the mechanisms are very different. The low frequency sounds would propagate omnidirectionally and would possess much more penetrative power than the high frequencies, due both to the much greater acoustic powers involved, and because of the wavelengths being very long compared to the wall thicknesses. The first internal lining of the rooms (behind the decorative fabric that is) is in this instance a kinetic barrier material (deadsheet) covered in a 2 cm layer of cotton-waste felt. The composite weighs 5 kg/m^2 and comes in rolls, 5 m × 1 m, and this is nailed to the stud framework of the room, felt-side to the room. Behind this barrier is an air cavity of 7.5 cm, containing a curtain of the same cotton-waste felt material, hung from the top, and cut carefully to fill completely the cross-section of the gap. On the other side of the stud frame is a double layer of 13 mm plasterboard, with a layer of 5 or 10 kg/m^2 deadsheet sandwiched in between.

All of the above layers are diaphragmatic; they are free to vibrate, but they are also all highly damped. The room, to the low frequencies at least, presents itself as a large, limp bag. When the sound waves strike the kinetic barrier, it is somewhat like the effect of a boxer striking a heavy sand bag: the room gives, absorbing much of the energy and converting it into heat. Effectively, the inner wall lining gets pushed and pulled around by the compression and rarefaction half-cycles of the sound waves, but its weight and internal viscous losses are such that it is almost incapable of springing back. Linings of this type have low elasticity, they are more or less inert, which is why such materials are known as deadsheets. In a similar way, you can give an almighty clout to a bell made from lead, but you will not get much ring from it. The lead will yield to the blow, and its high internal damping will absorb the impact energy. Its weight will then ensure that it does not move much, and hence if it can barely move or vibrate, it will have difficulty in radiating any sound.

When the sounds vibrate the deadsheet linings of our vocal room, work is done in moving the heavy, flexible mass, and acoustic energy is turned into heat energy as a result of the damping which resists the movement. Some sound is inevitably reflected back into the room, but with room dimensions so small, in a matter of a few milliseconds the reflected energy strikes another surface, so suffers losses once again. All in all, the sound decays very rapidly, and the low frequencies, below 150 Hz or so, are effectively gone in less than 100 ms. The *very* low frequencies receive no support at all from modal energy, as the pressure zone in a room of such size exists up to quite high frequency.

7.3.2 The pressure zone

Modal support was discussed in Chapter 1, and pressure zone concepts were outlined in Fig. 5.9, but let us look once again at the concept in a little more detail. If less than half a wavelength can exist within the dimensions of a room, instead of waves of positive and negative pressure distributing themselves over the room, the whole room will either be rising in pressure, or falling, depending on whether it is being subjected to a positive-going or negative-going portion of a long wavelength. The frequency below which the pressure zone will exist is given by the simple formula:

$$f_{pz} = \frac{c}{2\,Lr}$$

where: f_{pz} = pressure zone upper frequency
  c = speed of sound in metres per second
        (in studios, about 340 m/sec)
  Lr = longest room dimensions, in metres

In the case of our room under discussion here, the maximum room dimension is from the lower corner to the opposite upper corner, marked X and Y in Fig. 7.1(a), and is about 4 m, so the formula would give:

$$f_{pz} = \frac{340}{2 \times 4}$$

$$= \frac{340}{8}$$

$$f_{pz} = 42 \text{ Hz}$$

However, funnelling the wave into the corners with floppy walls and a diagonally placed felt layer is hardly representative of a reflective situation for a low frequency wave, so the longest side-to-side dimension would be more relevant. Substituting the 2.5 m side-to-side distance for the 4 m corner-to-corner distance, the formula would give a more reasonable pressure zone frequency as 67 Hz. There could therefore be no modal support in this room below 67 Hz, and the heavily damped 67 Hz mode would be the *only* supportable mode within the first octave or so of a bass guitar. The response can therefore be expected to be very uniform.

7.3.3 Wall losses

When this internal bag inflates and deflates, it does radiate some power outwardly, but the nature of the construction of the room is such that the air cavities created between the vertical studs and the outer and inner linings provide further damping on the inner bag. The air will tend to resist the pressure changes, as its elasticity will apply a restoring force on the inner bag. In turn, it will also exert a force on the outer composite layer of plasterboard and deadsheets. This layer is also very lossy and heavily damped. The internal particles in the plasterboard turn acoustic energy into heat by the friction of the particles rubbing together, and additional work is done by the need to move such a heavy mass. The sandwiched deadsheet forms a constrained layer, tightly trapped between two sheets of plasterboard. The constrained layer concept, shown diagrammatically in Fig. 1.5 and discussed previously in Chapter 1, effectively tries to sheer the trapped layer of viscously lossy material over its entire surface. The resistance to this sheering force is enormous, so the damping value is high, and the acoustic losses are great.

The transmission from the inner layer to the outer layers of the stud wall is partially due to the common studwork to which both surfaces are connected. This usually creates no problems as long as the inner surface is of a limp nature. Its lack of rigidity does not provide an effective acoustic coupling to the studs, and its weight, in turn, helps to damp the stud movement. The direct energy impact on the studwork itself, where the deadsheet is fixed to it, and hence where that deadsheet is effectively more rigid, is only a small proportion of the overall surface area. If the studs are 5 cm in width, and spaced on 60 cm fixing centres, then they occupy only 5 cm out of every 60 cm, or about 8% of the surface area. In addition, being of narrow section,

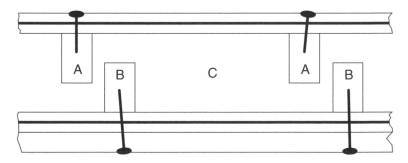

Figure 7.7 Staggered stud system. The sheet layers on each side of a stud wall can be attached to independent sets of studs, connected only by the top and bottom plates C. By staggering the studs and interleaving them, only a small amount of extra space is consumed compared with that of a conventional stud wall. This system reduces greatly the area of common coupling between the two sets of sheet materials, and usually offers more isolation than the more usual system with common studs

the lower frequency waves will tend to pass round the studs. However, if every last decibel of acoustic performance is required, the slightly more complex stud arrangement as shown in Fig. 7.7 can be used. Here, the studs are double in number, but are interleaved such that the only common couplings between the two surfaces of the wall are at the top and bottom, which are in any case coupled by a common floor and ceiling. Walls of this type take up an extra 2 cm of space, but if the extra performance is needed, the space penalty is low. Thinner studs could be used if the space was critical, but they would provide less damping on the panels, and hence may negate some of the advantage gained by the separation.

The majority of the inter wall-surface coupling is in any case through the air cavity. The cavity in this case is lined with cotton-waste felt, which helps to increase the losses, but at very low frequencies its effect is only minimal. The losses from the air cavity coupling are great as long as the outer surface is very heavy, as it is difficult for a thing of low mass to move a more massive one. It is therefore relatively difficult for the air which is trapped in the cavity to excite the outer composite layer of heavy sheet materials. Historically, some of the first tests of this principle were carried out by experimenting with cannons in the Alps, so as this factor of acoustic coupling is quite an important aspect of our isolation systems, perhaps we should now consider it in a little more detail.

7.3.4 Transfer of sound between high and low densities

In the 19th century, two cannon were placed on the side of a mountain, one low down, but not at the foot, and the other high up. The cannon were charged with equal amounts of powder, and observers were placed not only at each cannon position, but also high up and low down on a mountainside at the opposite side of the valley. The arrangement is shown in Fig. 7.8. The cannon low down was not placed at the very bottom of the valley, so that its sound would not be unfairly reinforced by having the valley floor to push against. When the cannon were fired, the flashes and the smoke were clearly seen from all distant stations, and as the distances to the listening stations were known, the sounds of the cannon were expected to be heard after the appropriate time intervals had elapsed.

The lower cannon was fired first, and the three listeners at the distant listening stations, A, B and D, awaited its report, which duly arrived at each position at the respectively appropriate times. In each case, when the sound was heard, the listeners signalled its arrival by means of flags. The sound intensity at each station was described as best as could be done in the days before sound level meters. The sound was heard loudly by the listener low down on the opposite side of the valley, and was clearly heard by the two listeners high up on the two sides of the valley, positions A and B in Fig. 7.8. When the higher cannon was fired from position A, again the flash and the smoke was clearly seen by the distant listeners, this time at positions B, C and D. After the expected time interval had elapsed, the listener at position B signalled that the report had clearly been heard, but after more than enough time had passed for the sound to arrive at positions C and D, no signals were seen, as no sound had been heard.

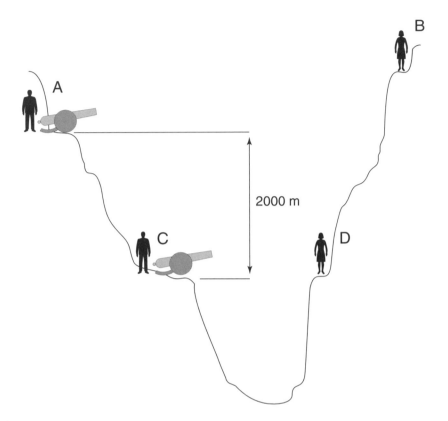

Figure 7.8 Cannon experiment. Two identical cannon are positioned at A and C, and charged with equal amounts of powder. When cannon C is fired, the observers at A, B and D will see the flash almost instantaneously. After a length of time corresponding to the local speed of sound in the air, and their distance from C, the three observers will hear the report from the cannon. When the cannon at A is fired, the observers at B, C and D will see the flash. After the appropriate time lapse, the observer at B will hear the sound, but the observers at C and D, in the denser air, may hear nothing, despite the fact that they are closer to A than is observer B

From this it was deduced that the density of the air in which the sound was produced, relative to the density of the sound in which it was heard, was responsible for the efficiency with which the sound would propagate. The explosion in the higher density air, low down in the valley, could easily cause the sound to propagate not only to the listening station D, in the same high density air, but also to the listening stations at A and B, higher up in the lower density air. However, whilst the report from the high cannon could be readily heard at listening position B, which was in the same low density air, the sound could not penetrate effectively into the higher density air at the lower listening stations, C and D. This was despite the fact that position C was closer to A than was position B, where the report was clearly heard. Furthermore, compared to the denser air at the bottom of the valley, the thinner air higher up the mountain provided the explosion of the gunpowder with less air to push against as it left the barrel of the cannon. With less air to push against, less work could be done by the explosion, and hence less work done meant less sound generated. The air pressure reduces by almost 1 millibar for every 8 metres that one rises above sea level, and if temperatures are equal, the densities will thus be proportionately reduced. In fact, given the 2000 m vertical separation between the cannons shown in Fig. 7.8, the air pressure at the upper positions, A and B, would be less than 75% of that at the lower stations, C and D.

7.3.5 The weight of air

As we are surrounded by air from the day that we are born to the day that we die, there is a tendency for many of us to take it for granted, but to the musical acoustician it is the prime medium for consideration. Without understanding its properties, the science of acoustics cannot be applied, so now that we have considered its density, perhaps we should consider its weight. In fact, the weight of air surprises many people who have not studied the subject. When a jumbo-jet flies at an altitude of 10,000 m, the inside will usually be pressurised to the equivalent of about 2000 or 2500 m. Above an equivalent pressure altitude of around 3500 m, the air is too rarefied to provide enough oxygen for human beings to breathe and remain conscious; unless, that is, one's blood has had some time to acclimatise and change its constitution (hence high-altitude training by athletes to enrich their blood before competitions). For this reason, the pressure in the aircraft *must* be kept below the equivalent pressure of 3500 m, unless oxygen masks are worn. If passenger comfort only were to be considered, it would be better to pressurise the aircraft to the sea-level pressure, as ear-popping, dry skin, thirst, and other side-effects of depressurisation could be avoided, but there are two reasons why this is not done. The first reason is that by reducing the pressure inside the aircraft, the pressure differential between the inside and outside is reduced, and hence the stresses on the air-frame are reduced. It is thus not necessary to build the aircraft so strongly if they only pressurized to 2500 m, and the weight saving from a lighter air-frame allows less fuel consumption, and hence reduces the running costs and ticket prices. The second reason however, is less obvious. The difference in weight between a jumbo-jet pressurised to sea-level, and one pressurised to a pressure altitude of 2500 m, is about half a ton. Transporting that half ton around for the life of the aircraft would cost a fortune in extra fuel. When one thinks about it, a compressed air cylinder for scuba diving is considerably heavier when full than when empty, and one can fill an awful lot of compressed air cylinders from the air contained within a jumbo-jet.

I hope that the descriptions in the last few paragraphs, highlighting the very considerable presence of air as a fluid, will aid in the understanding of the effect of the air gaps between the different layers of acoustic wall structures. They are certainly not the empty spaces that they are often thought to be. If the physical properties of the air are duly taken into account in the consideration of the wall structure, they can be used to benefit, not only in the acoustic damping of the wall systems, but also by creating more losses due to both the effects of the density *and* the acoustic impedance changes which the air can provide.

7.4 Combined effects of losses

After this long detour into the physics and acoustical properties of air, let us return again to our room. It can now be appreciated that the acoustic energy will be severely attenuated by the compounding of the low frequency losses in the deadsheet, the damping loaded upon it by the air in the inter-stud cavities, and the mass of the plasterboard to be moved; damped by the sandwiched, heavy deadsheet. As this whole structure is set on a layer of foam of an appropriate density, it thus sits on a foam spring, surrounded on the other five sides by air springs. The radiation of what energy is left must now pass through the 'springs' in order to get to the plasterboard layer which covers the foam, which is in turn glued to the structural walls. Springs are reactive; that is they tend to store and release energy, rather than passing it on. The cavity in Fig. 7.1 is lined with cotton-waste felt. Any resonances which try to build up in these air cavities, particularly as lateral movement of the air, or waves travelling around the box, must effectively pass through metres of felt, which they patently cannot do. By such means, resonances in the air space are effectively prevented. One acoustic 'short circuit' is therefore avoided, as if the air was resonating, its ability to couple the two sides of the gap would increase greatly.

The low-density air surrounding the inner, floated, acoustic shell must then transfer its vibrational energy to the much heavier plasterboard wall linings. As with our cannon, it is difficult for a low mass of low-density material to excite a high mass of high-density material, so further transmission losses take place as the mass of the plasterboard resists the movement of the air in contact with it. The particles in the plasterboard also provide losses due to their friction, so this reduces even more the energy which can be imparted to the foam to which it is stuck, and which in turn is stuck to the wall.

The foam itself, securely bonded to the plasterboard by adhesive, strongly resists the movement of the plasterboard, so yet again, vibrational damping is provided and even more acoustic losses take place. Finally, the foam, glued to the structural wall, flexes under the force being imparted to it from the plasterboard, because its mass and stiffness are insignificant compared to those of the main structural wall which it is trying to move. The exact effect of this final loss depends mainly on the mass of the wall to which it is attached, but typically, from the inside of the finished room to the other side of a structural wall, 60 to 80 dB of isolation at 40 Hz could be expected from a structure similar to that being discussed here.

Of course, what we have been considering here on our journey through the complex wall structure are transmission losses, but we should not forget that from each layer boundary, and particularly the inner boundaries of the more massive layers, acoustic energy will be reflected back towards the room. We have four heavy layers in this structure, the deadsheet inner lining, the composite plasterboard/deadsheet sandwich on the outside of the inner box, the plasterboard which is glued to the foam, and the structural wall. We discussed earlier the progress of the reflected wave from the inner lining, but as we pass to the sandwich layer, the reflexions will be stronger due to the greater rigidity of the material. However, these reflexions cannot pass directly into the room. They must first pass through the inner deadsheet layer, which has already attenuated the incident wave by its mass, its internal viscous losses, and its poor radiating efficiency. The reflexions from the sandwich layer will also suffer similar losses as they seek to pass through on their way back into the room. What is more though, just as some energy of the incident wave is reflected back into the room from the internal lining of deadsheet, the lining will also reflect back the reflected wave into the wall structure, towards the sandwich, a portion of the energy reflected from that sandwich, so round and round we go, losing energy at all stages. The more internal losses which we can create within our multi-layered wall by using internal reflexions to trap the sound within the layers until it dissipates as heat, then the more pure absorption (i.e. not counting transmission) we will be able to achieve.

Similarly, the other heavy boundary layers on either side of our 6-cm foam, stuck to the structural wall, will also reflect energy back towards the room, but the deeper we go through our complex wall structures, the more difficult it will be for the acoustic energy to find its way back into the room. Taken as a whole, such a complex wall structure as in Fig. 7.1 deals with our internal acoustic problems *and* our sound isolation problems by its high overall internal absorption. The structure is progressively absorbent, in that the most reflective layers are those furthest away from the inside of the room. If we remember our open window and brick wall discussion, simple sound-absorbing materials tend to be poor isolators, and good isolators tend to be poor absorbers. In the case of our vocal room, though, we need both good absorption *and* isolation. The option of placing an adequate amount of a sound-absorbing material on the inside of a room, then placing a simple iso-lation wall outside of it, tends to use enormous amounts of space. Using such simplistic techniques in a space of 3 m^3 (the same as the structural space for the room in Fig. 7.1), we would end up with a usable space only about the size of a telephone kiosk.

The multi-layered, complex structures described here are much more effi-cient in terms of the space which they consume. They are also a little like the energy absorbing front ends of motor cars which crumple gradually on a frontal impact, absorbing the energy of a crash stage by stage, and prevent-ing it from doing too much damage to its occupants, or to the occupants of the car which it may have run into. Similar principles are now employed in the new, lightweight armour for battle tanks and warships, in place of the enormously heavy, steel armour-plate which was used, almost exclusively, up to the late 1960s. The similarities here should not be too surprising as, after all, energy absorption is energy absorption, whether it is absorbing the energy of car impacts, shells, or sound waves.

Well, now that we have dealt with the relatively simple surface treatments which absorb the upper-mid and the high frequencies, and we have discussed the limp bag which so effectively controls the low frequencies, all that we are left with are the low-mid frequencies. These are the modal frequencies, where resonances can develop between the faces of the room surfaces. Their waves are too long to be absorbed by the superficial layers of absorbent felt, but do not carry enough energy to punch their way through the deadsheets and composite layers. In the modal controlled region, there are two types of mode – the forced modes, which are driven by the sound source but which stop when the forcing stops, and there are the resonant modes, which can continue to ring on at their natural frequency long after the driving force has ceased. A loudspeaker, when driven by an amplifier, will vibrate at the driving frequencies, but when the drive is disconnected, and if there is little damping, the cone will continue to resonate for a short time at its natural frequencies. The mechanism in the cases of the room and the loudspeaker is similar.

The problem in terms of the decay time of the room is the resonant modes, the worst offenders of which usually develop between parallel surfaces or opposing corners. A discussion of modal patterns was given in Chapter 1. In the room of Fig. 7.1, no parallel surfaces exist, as the walls have been angled in order to use geometrical techniques to help to prevent modal reinforcement. The amount of angling shown would not be enough in a more lively room, as the lower of the lower mid-frequencies would still see them as relatively parallel in terms of their wavelength (see also Chapter 2), but in the case of a room lined with so much felt and deadsheet, the gradual, non-perpendicular impact tends to suffer much greater absorption and much less effective re-radiation than would be the case when striking similarly angled, but more reflective, surfaces.

The ceiling structure of our vocal room is similar to that shown in Fig. 2.2(b). It is vaulted with the PKB2 composite deadsheet/felt material, the felt side facing the floor. The 20-cm depth of the vaulting is sufficient to begin to take the energy out of both the direct mid-frequency waves and the ones reflected almost perfectly from the hard floor. As the wooden floor is a better reflecting surface than any of the other surfaces of the room, the design of the ceiling opposing it has to give due consideration to that fact. Above the inner ceiling is a void of about 30 cm depth. When stuffed with absorbent material, be it fibrous or off-cuts of foam, it provides good absorption of any lower mid-frequencies which may penetrate the inner linings. The ceiling void is actually there primarily for another purpose however; the passage of ventilation ducts, as one does tend to have to provide vocalists with a good supply of fresh air, delivered as smoothly and silently as possible. Nonetheless, the area provides useful space for extra absorbent material to add even further control to the room.

7.4.1 A micro-problem

In small vocal rooms, subjectively, the lower mid-frequencies tend to be the most difficult ones to suppress, but in a room such as the one which we are discussing here, the remnants of modal energy which may be detectable after making a sudden, loud 'a' sound, as in *cat*, tend to be no problem in practice. However, as I learned the hard way, these rooms are better assessed through microphones than by listening inside of them. When I completed one of my first rooms of this type, although quite a bit larger than this one and with an extra window, there was a resonance which was detectable at about 400 Hz, and which lingered for a short time after impulsive excitation of the room. In spare hours, I searched for days for ways to control the problem but it simply refused to go away. The studio finally went into use during the time that we were building a second studio for the client, but I thought it a little strange that we had heard no complaints from the recording engineers about the vocal room resonance. In fact, the only comments about vocals recorded in the room were to the effect that they were beautifully clear. So, without drawing attention to the problem, I asked to hear some of the vocal recordings made in the room, and sure enough, I could hear nothing of the problem resonance, not even with the vocal tracks soloed. The room had a sliding glass door which faced the control room, and a window in an adjacent wall for visibility through to the main studio room. Though this wall was at 90° to the wall with the doors, the window was raked back about 8° from the vertical. There was also a very small window leading to the small vocal booth of the adjacent studio, but this faced directly towards neither of the other windows. The floor was of ceramic flooring tiles. Some thing or things in this room conspired to produce the resonance, but we never found the cause: finishing the other studio was a more pressing task. We conjectured that the recordings were clean because the directionality of the microphones when pointing in the usual direction did not face the source of the problem, which was no doubt partly true.

Eventually, more out of interest than necessity, we brought in equipment to hunt for the rogue resonance, and we were all surprised to find that it was about 70 dB down on the direct sound. Apart from the life provided by the single reflexions from the windows, glass doors and the floor, all of which were 'one shots', there was almost nothing left in the room response except for our 'problem', which could perhaps have been a cavity resonance between the double-glazed windows, but we soon forgot about it. It is strange how something can be magnified in one's imagination if one knows a problem to be there. Once I knew that this problem was not a problem in practice, I rarely noticed it again, except if I listened purposely for it. This story does go to show, though, just how effective the treatments described in this chapter can be in taking the decay problems out of a small room, yet allowing sufficient single reflexions to give a feeling of reasonable life and rendering the room pleasant to be in. We often notice this type of effect on a larger scale when putting the first acoustic treatment in buildings. It is surprising how many unpleasant sounds are exposed as the process takes effect before each one is finally eliminated. These sounds are all there in the original room sound, but many only become apparent as other room effects are removed. It is a little like peeling the skins off onions, where, as one is removed, another, smaller, different one becomes apparent. In very low decay-time rooms it is quite possible to begin going round in ever decreasing circles, trying to remove every last trace of the room's artefacts. It takes some experience to know where to stop, and the final recorded sound quality is the ultimate arbiter.

7.5 Trims

It is customary in such rooms to provide skirting boards to protect the walls when cleaning the floor, and also shin rails, on which to mount sockets for microphone inputs and power outlets. A dado rail is normally also provided at waist height, to give people something to lean on and prevent them from falling through the fabric. If these are only 10–15 cm in height, as depicted in Fig. 2.17, the specular reflexions that they may provide, even if the surfaces are varnished wood, seem to be of no subjective consequence, as they cannot support resonances. Frequencies below around 2 kHz just pass round them as a river flows round a projecting rock. Fabrics, though, *can* be reflective if they are too stiff. Fabrics such as calico can be so tight when stretched on their frames that they ring like drum skins. The best fabrics to use are similar to the ones used in the neutral rooms of Chapter 2, soft ones with a relatively open weave, or 'stretch' type dress fabrics, such as Lycra, which are also excellent. They remain neatly taught, and spring back if anybody pokes them, yet they remain limp enough and are sufficiently open to be sonically inert. There are also a number of manufacturers now making fabrics specifically for acoustic purposes, and these are often flame retardant, or even totally non-flammable, for use in public spaces such as theatres and concert halls.

7.6 Conclusion

I find rooms such as these *very* effective for vocal performances, and also very useful for getting tight, powerful bass guitar recordings, but for other instruments they are not too inspiring. On the other hand, I know people who do like to get 'clean' recordings of instruments so that they have the maximum potential for processing in the post-recording stages, and for such purposes these rooms are excellent. One can also experiment with pointing the microphones at the floor or the glass doors, as some interesting effects *can* be achieved, but if such effects are required, it is simple to install reflective panels when needed. All in all though, for vocal rooms, these constructional techniques work extremely well.

Chapter 8

Room combinations and operational considerations

As has no doubt become clear by now, no one room can perform all the functions which may be required for the recording of the whole range of probable musical performances. The large, variable room is perhaps the best option, if money and space allow such a room to be built, but even the best such room can never mimic the performance of a good, small live room, or a good stone room: at least not without absurd complexity of structure and mechanisms. We should now therefore consider how we could put together the pieces so far discussed in order to make a viable, practicable, and efficiently flexible studio package.

8.1 Options and influences

Let us imagine that we were presented with a building in which to construct a studio, and that we had the luxury of 500 m² of floor space and 6 m of ceiling height. One option would be to sub-divide it as shown in Fig. 8.1, which would give us enough space to construct a large, variable, general recording area, plus a stone room, a moderately live room, and a vocal/dead room. However, the layout of such rooms is full of compromise, forced upon us by many conflicting priorities. Ideally, everybody needs to be able to see the control room, but also, everybody needs to be able to see each other. At the same time, the general influence of the positioning of the smaller rooms should not detract from the optimum shape of the main studio area. This may all seem rather obvious, yet it is surprising just how many studios are built without even the most basic of these requirements being given their due consideration – this even includes some studios which have ostensibly been professionally designed.

Figure 8.2 shows an actual situation that I was faced with, in a 90% completed form, when I was asked to act as consultant for the finishing of the job. Believe it or not, over $50,000 had already been spent on the rooms. The designs had been done by a theoretical acoustician with some theatrical acoustics experience, and not by a specialised studio designer, and all the rooms had been designed around the use of large semi-cylindrical diffusers, Helmholtz resonators, and absorbent/diffusive tiles made from gypsum and

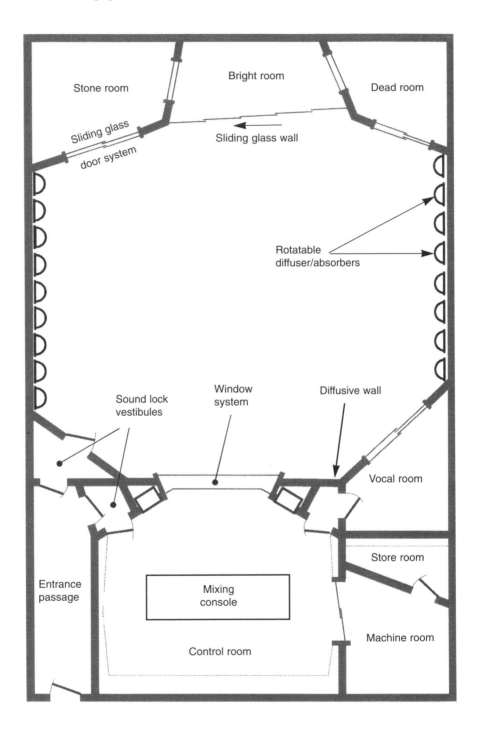

Figure 8.1 Layout of hypothetical 500 m studio, but based on an actual design

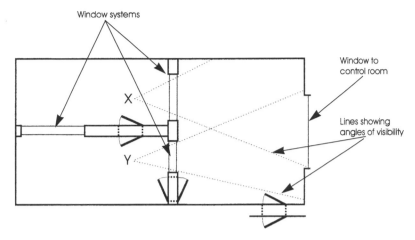

Figure 8.2 Studio designed by cinema acousticians. Musicians in positions X and Y, the natural positions for good visibility into the larger recording room and the control room, are unable to see each other

mineral wool. All the rooms had been designed to classic RT_{60} figures, and somewhat surprisingly, all had been aimed at a somewhat similar acoustic performance, except for the slightly longer RT in the larger room. The windows and doors had been situated in accordance with structural simplicity, rather than acoustical, operational, or musicians' needs. Perhaps this fact, more than any other, made the task of modifying this studio all but impossible.

I was too soft hearted to suggest to the owners that they should demolish everything and begin again, so I tried to rescue what I could of the original structure. After some days of sketching and discussion, it was ultimately the owners themselves who suggested a total re-build. What was so distressing

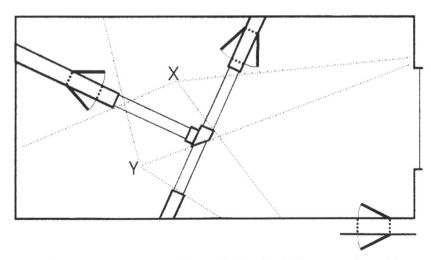

Figure 8.3 Layout which could have been workable using similar space and materials as those in Fig. 8.2

was that the simple re-positioning of the internal walls, doors and windows, such as shown in Fig. 8.3, it would have rendered the room much more rescuable. What is more, it would have cost no more to do so than to build what had actually been built. It seems however that the only design considerations had been separation and acoustic control. Neither of these things should be achieved at the sacrifice of the ease and comfort of the musicians, yet in this instance it turned out that the designer had failed, almost totally, to consider the needs of the musicians.

The sad thing about the above situation is that with a little more forethought and experience, the use of the layout of Fig. 8.3 would have allowed the construction of an excellent live room in section X and good vocal room in section Y, and the shape of the larger room would have allowed much better acoustic control, without having to resort to the 'battleship' methods which lost so much usable space in the original layout. Anyhow, now that the surprising ease with which some well educated people can get things wrong has been established, we can move on to take a look at things in a more orderly manner.

8.1.1 Demands from control rooms

Even 'external' things, such as the control rooms will make their demands on the design of the studio rooms. As I explained at great length in my book *Studio Monitoring Design*, whether you have a wonderful set of studio acoustics or not, it is like driving blind if the control-room monitoring and acoustics are not sufficiently sonically neutral and free from idiosyncrasies of their own. To make the best of what the studio acoustics, microphones, and musicians have to offer, it is imperative that the control room should not be compromised. The most likely compromise in this area is to visibility, which is indeed very important, but there are optimal locations in control rooms for loudspeaker positions, and I, personally, do not agree with the pushing of the monitor loudspeakers into high locations, where the ear perceives things very differently than from a horizontal direction, all for the 'luxury' of an enormous front window.

Large windows are problems in themselves. The bigger they get, the more they tend to allow sound through, or to resonate in unwanted frequency bands, unless they are of a very great weight, in which case they may need to be custom-made by a specialist glass factory, the cost of which can be alarming. Two-way closed-circuit television is one possible solution for visibility between the control room and any out of the way places, but there is normally the problem of the 10 or 15 kHz timebase whistles from the TV monitors. Some people find these to be extremely irritating, but perhaps more dangerously from a recording point of view, many people cannot hear them at all. In such cases, if they do pick up on the recording microphones, and especially if any top boost is applied on the console, then they can pass unnoticed onto the master recording, only to subsequently irritate the purchasers of the final product.

We also need to consider access to and from the control room, which in my designs usually means access via a side wall or the extreme sides or centre of the front wall. For me, the rear wall of a control room is sacrosanct, and should only be used for absorber systems. It is perhaps the most critical surface, as it takes the full impact of the incident wave from the monitor system, and anything coming back from that wall will colour the perception in the room. The low frequency absorber surfaces need to be as large as possible, so penetrating them with doors is generally undesirable. I now try to avoid the use of sliding glass doors in the centre of the front wall, even though I was one of the pioneers of the concept, at The Manor studio in Oxford, England in 1975. Unless the glass is *very* heavy, so much so that perhaps opening and closing the doors becomes a difficult exercise, any LF resonance which exists in the door system will not only cause a perceivable resonant overhang, but will also absorb some of the impact from the bass end of the monitor system.

If the control room is sufficiently wide, then a heavy window can be placed centrally, between the monitors, and heavy doors of composite nature can be located on the outer sides of the monitors. These can provide access to the studio rooms, and to the reception areas or access corridors if necessary. If I must provide side doors into a control room, I prefer them to be as far to the back of the room as possible, to avoid disturbing the absorption system on either side of the mixing position. However, if they must be at the sides, they can be angled so as to reflect any sound which strikes them into the rear absorber. Every situation is different, though, with its own set of specific demands, but that, of course, is what helps to make studio design so interesting.

8.2 Layout of rooms

In Fig. 8.1 the control room is flanked on one side by an access corridor, and on the other side, by a machine room and a vocal room. If any one of the studio rooms needs to be alongside the control room, it is better that it is the most acoustically dead of the rooms, as the SPL which develops in the live rooms can present much greater sound isolation problems if the live rooms are adjacent to the control room. Also, usually, the dead rooms will be used for much of the vocal overdubbing work. This process can consume a great deal of time, so a location close to the control room is convenient and expeditious. The arrangement shown allows access and visibility directly to and from the control room and the studio, so whether doing live vocals or overdubs, the vocalist will not feel too isolated.

So, already, having discussed so little about the studio layout itself, we have been diverted by considerations from some seemingly secondary sources, but their pull is great, and any studio designed without due consideration of the widest spheres of influence will likely fail to achieve its full potential. Different designers will have their own very special priorities which they will try to avoid compromising, and each of these points is likely to be based not only on knowledge of facts but also on insight gained through years of practical experience. This is largely what separates studio designers from other acousticians. In each case, the experiences will have been different, and the individual tastes of the designer will be different, so this will lead to some very different designs. This is actually quite fortunate, as without variety, things would be a little dull. Obviously it must be borne in mind that I, too, am a product of my circumstances, so a degree of personal leanings will be influencing what I write, and what I design. There is no bible which governs this subject absolutely.

8.2.1 Priorities and practice

Figure 8.1 shows a very attractive layout, but such a flexible option would be expensive to build. Studios are usually built on budgets that must be earned back over a matter of a few years, and so must be constructed with the appropriate economic considerations. When we cut our budgets, we usually also cut our options, so a designer must consider carefully the most likely uses to which the studio will most frequently be put. If it is proposed to record many voices, or chamber orchestras, then obviously a stone room and a live room would not be an appropriate choice of combination. In fact, such a combination would be unlikely to be a good choice in *any* circumstances that spring to mind. I do know of a few studios who earn their livings from a single stone room or live room, but they cater for a niche market. I am sure that if any of them were to build a second studio room, then it would either be a dead room, as described in Chapter 7, or a room with some degree of variability.

An interesting problem faced me recently. I was designing a studio for a company which had already built up a good base of clientele, but in order to continue in operation they had to rebuild, to improve the sound isolation and general acoustics. The studio had grown from a single studio to two studios, but the building was only converted into its original studio form with the one studio in mind. Studio businesses do not always develop as expected, and in this case they needed to be able to record a wide range of small musical groups, including acoustic, rock and electronic. They also did much voice-over work for foreign language replacement of television programmes and cartoons. The two new studios were required to operate either separately, or with the second control room relegated to being an editing room, and its studio room put into use with the main control room. The final layout is depicted in Fig. 8.4.

In this design, Control room 1 was the main control room for music recording. Previous experience had led the owner to believe that they needed a good live room, and a vocal room which would not colour the sound. Control room 2 would not be used for serious mixing purposes, but should be neutral enough to make a good editing room, and to be able to function as a control room for overdubs or dialogue replacement. In Fig. 8.4, it can be seen that Control room 2 has an adjoining studio (Studio 2), which is only large enough for one musician at a time, or perhaps a trio of backing vocalists, but this room also has to serve as a third recording space for Control room 1. The problem was that Studio 2 would best serve Control room 2 if it were a room which was as dead as the ones described in Chapter 7, but Control room 1 already had such a room adjacent to it, so doubling up on such rooms would seem to serve little purpose. It was eventually decided to make Studio 2 a bi-directional room, which could serve as a relatively dead room when being used with Control room 2, but would have more life when used with Studio 1.

It was presumed that when being used as an extension to Studio 1, the musicians would be facing Control room 1, and when working with Control room 2, they could face the opposite direction without losing visual contact with Control room 2. Figure 8.5 shows the idea in more detail. The main problem with this option was how to make a small room with the required brightness, but without the boxiness which usually accompanies such designs.

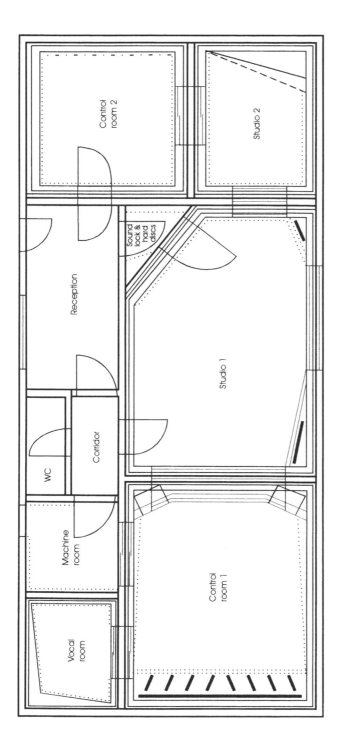

Figure 8.4 General layout of a small studio complex (Tcha Tcha Tcha Studios, Lisbon, Portugal)

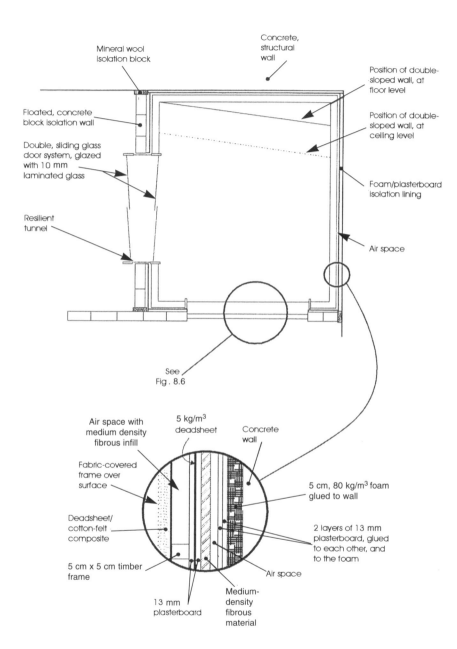

Mineral wool
isolation block

Concrete,
structural
wall

Position of double-
sloped wall, at
floor level

Floated, concrete
block isolation wall

Position of double-
sloped wall, at
ceiling level

Double, sliding glass
door system, glazed
with 10 mm
laminated glass

Foam/plasterboard
isolation lining

Resilient
tunnel

Air space

See
Fig. 8.6

Air space with
medium density
fibrous infill

5 kg/m³
deadsheet

Concrete
wall

Fabric-covered
frame over
surface

5 cm, 80 kg/m³ foam
glued to wall

Deadsheet/
cotton-felt
composite

2 layers of 13 mm
plasterboard, glued
to each other, and
to the foam

5 cm x 5 cm timber
frame

Air space

13 mm
plasterboard

Medium-
density
fibrous
material

Figure 8.5 Plan of small room using the double-sloped Geddes wall principle

8.3 Isolation considerations: doors and windows

Isolation was a great priority, not only because the two studios had to be able to work separately, but also because the poor reputation due to the bad separation between the old studios had to be ended once and for all. The wall between Studios 1 and 2 was a massive, sand-filled concrete block affair, floated away from the structural walls, floor and ceiling with high-density mineral wool. Each side of the wall was treated with 8 cm of 80 kg/m^3 polyurethane foam then by two sheets of plasterboard, all layers bonded with contact adhesive. All three studio rooms and both control rooms were separately floated on 120 kg/m^3 polyurethane foam (Arkobel) but Studio 2 had a second floated floor, of different density and thickness to the other floated floors. This was to avoid any common resonances in similar construction methods which could have reduced the isolation at any common resonant frequencies. The shell treatment of Studio 2 was essentially the same as that described in Chapter 7. The room began as a dead room, with the additional life of the double glass-door system to the control room, plus a quadruple-glazed window to Studio 1. In fact, before looking in detail at the system of livening Studio 2, perhaps we should first look here at the door and window systems.

8.3.1 Sliding doors

The two door systems were each made out of two panels, one fixed and one opening (sliding). The doors were mounted in the frames of each of the adjacent, floated boxes, whose walls were separated by a sand-filled concrete block wall. The non-parallel mounting of the doors helps to reduce the resonant energy in any modes in the cavity between them, and also allows them to avoid forming parallel surfaces with the opposing walls inside of their respective rooms. The glass chosen in this instance was 10 mm laminated glass, which is quite heavy, and also, by virtue of the laminating, very acoustically dead. The laminating layer acts as a constrained layer, in the same way that the deadsheet layer operates when sandwiched between the two sheets of plasterboard in the wall structures. The laminated glass also adds an extra element of safety as it is *very* strong, easily resisting the impact of wooden blocks thrown at it quite forcibly. There is little risk of people breaking it by accidentally walking into it, or knocking a guitar against it.

Between the two doors a tunnel was constructed, but was connected to one door frame rigidly, and to the other only by means of silicone rubber. The tunnel is therefore an extension of one room, but connects it resiliently with the other room. By this means, no significant direct transmission of sound can take place between the two rooms. The tunnel passes through the concrete wall, but it is spaced off with mineral wool or polyurethane foam to avoid direct contact with the isolation wall, and hence to prevent the transmission of sound into the main structure. Ideally, if conditions allow, the two doors should be of different widths, and the glass of different thickness, in order to reduce any transmission of sound via any common resonances, but one has to be careful here in applying rules too generally. Perhaps a 6 mm and 10 mm pair of glass doors would provide a better isolation than two 8 mm glass doors, which would contain the same total thickness of glass (16 mm), however, weight is an important factor, and experience has taught me to use the heaviest and most acoustically dead glass that I can find readily available. A 12 mm and an 8 mm glass would probably be *marginally* better than two 10 mm glasses, but the difference is rarely worth the extra cost of different door frames, and glass prices tend to rise disproportionately with thickness: 25% extra thickness can sometimes mean 100% more price. Varying the area of the panels is often just as effective as varying their thicknesses, and is often a less expensive solution.

8.3.2 Window systems

The window system between Studios 1 and 2 raises a number of interesting topics regarding window design. In this case, quadruple-glazing was used, with two panes mounted as widely apart as possible on the foam/plasterboard linings of the concrete centre wall, and one further pane on each of the floated boxes. The sound isolation between these rooms was more important than usual, because there would be times when the two studios were working independently, perhaps with a rock band in Studio 1, and a voice-over in Studio 2.

As explained earlier, the 20 cm, sand-filled concrete block wall was covered on each side with 8 cm of 80 kg/m^3 reconstituted foam and 2.5 cm

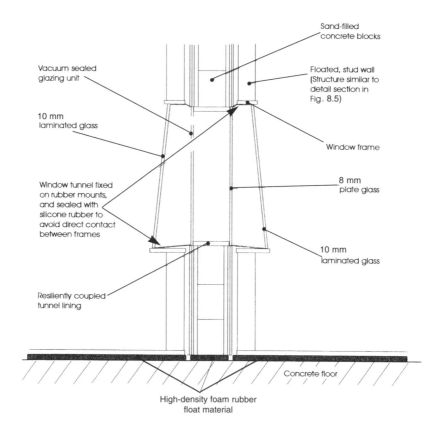

Figure 8.6 Quadruple-glazed window system

of plasterboard, which gave a total wall thickness of just over 40 cm. The hole in this wall was sufficiently large to allow as much visibility as necessary, but no larger, as the bigger the hole in the wall, the less the resulting isolation. The lines of sight were carefully assessed, and the minimum practical hole size was used for the windows. The central panes were different, one a vacuum sealed, double unit, and the other a pane of 8 mm plate. The windows in the walls of the floated boxes were correspondingly larger to allow the angular visibility, as shown in Fig. 8.6.

Again, as with the doors, tunnels run from the outer windows to the inner ones, fixed to the frames on the floating rooms, and attached by silicone rubber to the central windows. These are usually lined with carpet, to avoid chattering in the gap, and are normally made of a relatively acoustically dead material, of not too great a thickness, which will allow sound to be absorbed not only by the tunnel itself, but also by the soft materials packed around the tunnel. In this specific case, the outer panels of glass were of 10 mm and 12 mm laminated glass; different thicknesses being used because each was the same width and height. They were also angled such that they would reflect any sound striking them up into the absorbent ceilings of each of the rooms.

8.3.3 Multiple glazing considerations

In this instance, quadruple-glazing was used in order to reduce the sound transmission at each change in wall structure, and in particular to prevent the sound from reaching the concrete wall. Effectively, the inner pair of windows were on the plasterboard layers, which in turn were isolated from the concrete wall by polyurethane foam, but in many normal circumstances, say between two simple walls, a greater space between the windows can often be more effective than more windows. In other words, if the space between two walls is 80 cm, then it *can* be more effective to use two panes of glass of 10 mm and 12 mm, than to split the tunnel into spaces of 20 cm, 35 cm and 25 cm by the use of four panes of mixed 10 mm and 12 mm glass. It is possible that the mid and high frequency isolation of the quadruple-glazing would be better, but that the low frequency isolation would be worse, as the space between the panes is an important factor in low frequency isolation. There are few absolute, hard and fast rules to this, as the wall structures, the degree of possible angling of the glass panes, and the type of work in which the studio is usually engaged will all add their own special conditions. For example, a voice-over studio, or a talk studio, is unlikely to be placing great importance on very low frequency isolation, as low frequency sounds are not likely to exist at high level on either side of the window. In this case, the quadruple-glazing option may be preferred, but for a studio mainly producing dance music, with lots of low frequencies, then quite possibly the double-glazing larger space option would be preferable. Each case must be assessed individually.

8.3.4 High degrees of isolation

There is obviously no need to produce isolation in a window system that is any greater than the isolation provided by the walls which they penetrate. A 70 dB window system in a 55 dB wall system only succeeds in wasting money. However, if *very* high isolation window systems are required, they may become absurdly expensive. Eastlake Audio installed a window system[1] in Belgium, where an isolation of 80 dB was required between the control room and the studio. The glass panes were 11 cm thick (yes, centimetres!), and weighed almost 1 ton each. The thickness was the maximum that the particular glass-making machines could stand the weight of, and the transport was very difficult. The original specification called for 14 cm glass, but when it was found that it could not be acquired in Europe, panes of 11 cm were used with a greater space between them than had originally been planned. The cost of all of this I dread to even contemplate. In fact, the total weight of the two rooms, floated on steel springs, is almost 2000 tons, with 9 tons of rubber in the walls. It is certainly impressive, but such would hardly be practical solutions in most cases. In fact the following quotation from David Hawkins, the owner of Eastlake Audio, just about sums up the reality: 'Normally, when people ask for isolation figures of this type (80–85 dB), I generally suggest that they build the rooms in different streets.'

Conventional, hinged doors also pose their problems however, doors are usually able to be positioned in areas of the structure which are less critical than the normal window locations. What is more, right-angled bends and isolation vestibules can often be incorporated into the designs of door systems. In general, doors need to be heavy, acoustically lossy, and well sealed into their frames, but perhaps an extreme case can once again be outlined by describing the doors at the Belgian studio mentioned above. They used industrial doors with a nominal isolation value of 55 dB, which were employed either in pairs, or in triple sets between critical areas. Each door weighed 300 kg, and was suspended on ball-bearing hinges. When the closing handles are turned, they apply a pressure of 400 kg on the door seals, and hence on the walls. The repercussion of this is that the structure of the walls must be very strong and heavy, which implies more cost, and places restrictions on where they can be used. Not in a third-floor domestic bedroom studio, for example.

It is often difficult to get the fact across to people that 70 or 80 dB of isolation demands the same degree of treatment, irrespective of the purpose for which it is needed. 'But I only want it for practising my drums' is a cry so frequently heard. It is as though people expect that the cost of isolation somehow reflects the cost of what they are doing. Because they are only spending the cost of a drum kit, they expect that the isolation to the bedroom of a neighbour will somehow be cheaper than if they were spending a hundred times as much on a complete set of studio equipment. Obviously (or at least it *should* be obvious) the two situations will have an identical isolation requirement, whether the drums are used for practice only, or for a very serious recording process. Basically there are only two factors involved in extreme cases of sound isolation: great weight or great distance. This brings us nicely back to the consideration of the layout of rooms in multi-room studios, as the isolation between sections can be destroyed by inappropriate door and window systems. Anyhow, now that we have dealt with some of the door and window considerations, we can return to other aspects of the acoustic design of Fig. 8.5.

8.4 The Geddes approach

In Studio 2 of the layout which we were discussing before our diversion into door and window systems, the problem was how to introduce life into the

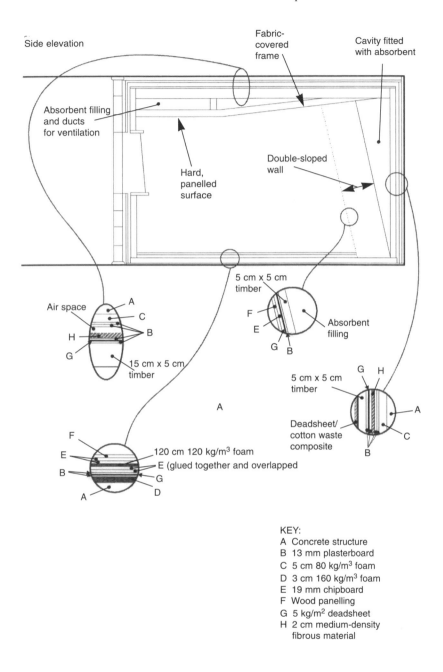

Figure 8.7 Isolation and acoustic treatment details of a small 'Geddes' room

room, without creating any boxiness in the recorded sound. Figure 8.5 shows a floor plan, with the dotted line showing the line where the angled wall strikes the ceiling. Figure 8.7 shows a side elevation of the room. From these figures it can be seen that the wall opposite to Studio 1 is a double-sloped affair. I first came across this concept when hearing of the work of Dr Earl Geddes[2], who proposed a double-sloped wall in an otherwise rectangular room for the more even distribution of modal energy in control room shells. Well, I have reservations about that particular use[3], but the technique seemed to provide a useful solution to some of the more intransigent small recording room problems.

Essentially, the double slope allows a relatively steeply angled wall surface without undue loss of floor space, although this is all in relation to the room size. The steep angling tends rapidly to steer the modal energy into the oblique form, which passes round all six room surfaces. Remember, on each contact with a wall, a sound wave will loose energy by absorption, and that absorption tends in many cases to be greater for waves striking the surfaces at oblique angles than for perpendicular impacts. By making all surfaces absorbent, other than the glass, the wooden floor, and the double-angled wall, the modal energy can be suppressed very quickly.

Figure 8.8 shows the room in typical use with Control room 2. In this situation, voice replacement work is being done. The person speaking, or singing, is facing the double-sloped wall. The wall is panelled with wood strips to add life to the room, and into it has been recessed a monitor television for voice synchronisation. The whole wall is hard, but only little reflected energy reaches the vocal microphone, as the window behind the vocalist is angled steeply upwards into the absorbent ceiling. What is more, when the room is being used in this way, it is often desirable to draw curtains across the window. This gives more privacy between the two studios, and the curtain, if heavy, and deeply folded, also helps to suppress reflexions. The double-angled wall is of a relatively lightweight construction, and the whole

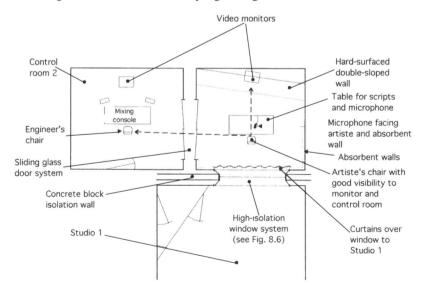

Figure 8.8 Studio 2 in use for voice-overs

cavity behind it is filled with mineral wool and scraps of felt and deadsheet, which press against the rear of the wall to provide considerable damping. The cavity itself is heavily damped by the filling.

If we now consider this room, one frequency band at a time, we will see how the overall design is effective for wideband control. Due to the overall geometry, the mid and high frequencies can find no simple, effective, reflective path back to the microphone. The upper and mid-range of the low frequencies, for which the angling of the window may be insufficient, do not 'see' much of the sloping wall, but either pass through into the great depth of absorbent filling, or are absorbed by the highly damped nature of the wall structure itself. The very low frequencies are in the pressure zone of the room, so are in any case rapidly lost. The pressure zone frequency for this room is about 65 Hz. Visibility between the engineer and vocalist is good. Each can face their own monitors, and with just a small turn of the head, can see each other clearly.

In the case of the room being used in conjunction with Studio 1, the musician(s), in this case let us presume that it is a drummer, will face away from the double-sloped wall. This will give direct visibility to Studio 1, and on through to Control room 1. There is also excellent visibility to Control room 2, in the event that it should be being used as an additional isolation room. This situation is shown in Fig. 8.9, and it can be seen that the microphones around the drum kit are generally pointing backwards and downwards. They are thus pointing not only at the drums, but on towards the reflective floor and double-sloped wall, and hence will pick up a great deal of early reflected

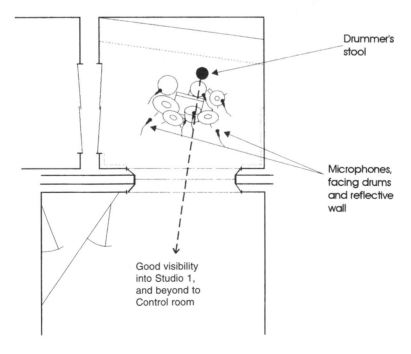

Figure 8.9 Studio 2 in use in conjunction with Studio 1

energy which will 'fatten' the sound of the drums. The drum kit will not sound the same as a kit recorded in a live room, nor of one recorded in a large room, but it will certainly, for almost all purposes, be preferable to the recording of a kit in either a more conventional small room or in a dead room. There will be a more potent drum sound and more 'feel' for the drummer than would be the case in a small dead room, and only very low levels of unnatural room colouration would result. The room does not possess the small-room boxiness which spoils so many recordings, and in fact, after the first year of use, the studio's engineers reported that this room was both very well liked and very well used.

Figure 8.10 Studio room, live one end, dead the other (Tcha Tcha Tcha Studios, Lisbon, Portugal)

8.5 Recording techniques for limited acoustics

In the small studio complex depicted in Fig. 8.4, the main recording area was relatively live, having windows to Control room 1 and to Studio 2 on two of its sides (its two ends), and a large window to the outside on another side. This wall was of angled brickwork installed in a sawtooth arrangement, see Fig. 8.10, and faced a dead wall on the other long side of the room. The floor was of wood, over which rugs could be laid if required, and the ceiling was V-shaped, being hard at the side nearest to the control room, and soft at the Studio 2 side. The room contained a grand piano, the open lid of which could face the hard surfaces if a rich tone was called for, or could face the absorbent 'trap' wall if more separation was needed, such as when recording a jazz quartet or similar.

In this room, other than by laying down rugs or using movable acoustic screens, there was little variability provided, as the owners deemed the range of rooms that they had to be sufficient for all of their normal needs. Because they had insisted on such high degrees of sound isolation, in which weight *and* space were consumed, they did not want to lose any further space by the installation of systems of variable acoustics. Subsequently, however, the ceiling of the live area was made to be more absorbent, because in the state shown in Fig. 8.10, it proved to be, not surprisingly, excessively lively. Nonetheless, skilful use of the spaces is now rendering good recordings, and there *are* ways to get around the limitation of fixed acoustics.

8.5.1 Optimum use of space

In Section 3.2.1, I referred to a recording which involved the replacement of five acoustic guitars on a live TV recording, destined later for CD release, though not originally intended for such purposes. In the replacement process, in a relatively live room, all the guitars were recorded sequentially in the same spot in the room, and although the room sound was not unduly evident on each track, when the five guitars were mixed together, the room sound was rather too much, certainly for *my* liking. There is, however, an interesting way of avoiding this, which involves moving the guitars to different room locations, so that each one carries a different ambience, not only due to the collection by the microphone of different reflexions, but also because the room is *driven* differently, from different nodal and antinodal points. This way, not all of the same room resonances are generated, so the tendency for an overpowering build-up of any particular resonance is reduced. In fact by moving the instruments *and* the microphones in this way, each will add its own variability to the sound. Nonetheless, whilst this can avoid an unnecessary build-up of resonances and can be a very effective technique, it still adds to the phase confusion, characteristic of the use of multiple microphones in ambient spaces. In exposed cases though, if something more akin to a simultaneous, stereo-pair recording is required, there is an interesting way to simulate this.

8.5.2 Moving musicians

There is a technique for multi-tracking vocals which I first heard of when in the USA. Its principles are so obvious when you know about them, but there again, such is the case with many things. Figure 8.11 shows a stereo pair of microphones in a room with an appropriate ambience. If it is desired that two vocalists build up the sound of ten, by means of five unison or harmonised recordings, then instead of grouping them around a single microphone and panning the different tracks into different locations on the stereo mix, a single stereo pair of microphones can be positioned at the chosen spot in the room. The vocalists then move into different locations for each take. If ten recording tracks are available for this, each stereo pair would be panned left and right in the mix, or to whatever other desired position, and the spread of panning would be built up automatically. Usually, the sound of such recordings is much more spacious, powerful, and natural than the 'five panned mono tracks' approach. If ten tracks are not available, then, especially if digital recorders are used and noise build-up and generation loss are not problems, each subsequent recording can be mixed with the playback of the previous recordings, carefully balanced of course, and recorded onto another

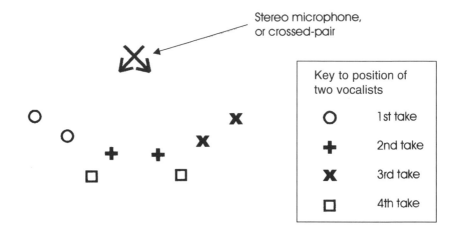

Figure 8.11 Method of creating a more natural ambience when multi-tracking backing vocals. This technique can create a better choir effect than merely panning voices which have been recorded in a fixed position

pair of tracks. The previous pair of recorded tracks can then be used for the next 'bounce', and so forth. This way, only four tracks are used, and anything which may be needed to 'repair' the final build-up such as earlier takes which are now considered to need reinforcing, can be recorded separately on the last but one pair, and either bounced again, or left separate until the final mixdown. There is no extra ambient build-up by the use of this technique, as no extra energy is driven into the room by recording five pairs of vocalists separately than would be supplied by ten vocalists singing simultaneously. It is a technique worth trying, believe me!

8.5.3 Changing microphones *and* positions

Techniques such as these can greatly improve the flexibility of rooms in which the quantity of acoustic variability is not great. Another possibility is to change microphones for different overdubs. If other microphones can be used which are still considered to be acceptable sounding for whatever is to be recorded, then as each microphone type has a distinctive and different polar pattern/frequency characteristic, the build-up of the room artefacts, by multiple recordings in the same place, can be significantly reduced. There are limits however.

8.5.4 Phase

I recall in the mid-1970s being asked to record David Bedford and Mike Oldfield, playing organ and guitar respectively, in Worcester Cathedral in England. This was only four days before the old organ was due to be dismantled. Despite being the technical director of The Manor Mobiles at the time, I used the Rolling Stones' mobile for the event, as The Manor units had other work on that day. (It was Virgin Records' policy never to turn down an outside client for the sake of an in-house project. Your own work could be ordered to return to your own studios, but if an outside client went to another company because you could not accommodate them, then they may stay with the other company for future work. This was considered to be bad for business.) Anyhow, the Rolling Stones' mobile had good staff, good sounding equipment, and an excellent set of microphones, so I felt no loss in being deprived of my own units.

Mick McKenna and I spent a great deal of time, during the rehearsals, walking round the cathedral, listening to the majesty of the sound in different locations, then discussing the most appropriate microphones to use for each location. All in all, we positioned 22 microphones in different locations, hoping to take best advantage of the sounds of each of the sections of organ pipes in the natural reverberation of the cathedral. Each microphone was fed to a different track of the 3M 24-track machines, and once each channel had been assessed for its individual sound during the afternoon rehearsal, Mick stayed in the truck, looking after levels, and I went into the cathedral to savour the atmosphere. As it went dark, the only light in the cathedral was from two candles, one on the organ console, and one on the 100-watt Marshall guitar stack at the opposite end of the cathedral. The music was appropriately named 'Instructions for Angels'.

It soon became very eerie, but as I wandered around in the dark, this somehow intensified the music. I kept getting frights when approaching the many life-size statues dotted around the cathedral, but finally found a comfortable stone block on which to sit whilst I absorbed the music, though still with a prevailing sense of unease. Eventually, we were thrown out by the police. It was getting very late and it appeared that we were keeping half of Worcester awake. The sound isolation in these buildings is not what it could be! When the dim electric lights came on as the police arrived, I had another sudden fright when I found that I was sitting on the tomb of King John: perhaps something that Robin Hood would have rather enjoyed rather more than I did.

The following week, I went to the private studio of Mike Oldfield to begin mixing. The album was scheduled as Virgin's first release using the BBC 'Matrix H' quadraphonic coding system, and Mike's studio had a set of Eastlake Audio, quadraphonic monitors. I listened to the various tracks of the multi-track, pair by pair in the front or rear loudspeakers, or, at times, a pair on each front/back side. It was really impressive and I played around for hours. Time was of no consequence; it was a private studio and was not being charged by the hour. Eventually, out of the 22 microphones recorded, I decided to use about ten. Then, horror struck.

When more than three or four microphones were brought into the mix, they were usually destructive. The phase relationships simply would not add. Nor, in fact, would they subtract overall. They were entirely random, and no amount of phase changing could make things work. Of course, I knew this in principle and immediately recognised what was happening, but I had never had the problem quite so startlingly presented before me. In stereo, the destruction was even worse, and it was so intensely frustrating because I knew what amazing sounds I could imagine by listening pair by pair, or even with pairs of pairs. In the end, many of the separate microphones *were* put to good use, as for example when the organist changed from the choir organ to the swell organ, or the great organ to the main. Edits could be made in the mix, changing at the appropriate time to pairs of microphones considered to be most rich or revealing for each section of the music. Unfortunately though, the magic sounds that I could build in my imagination could not be built acoustically.

I knew exactly *why* I could not have what I wanted in the mixing of the microphones, and I knew that it was pointless even *trying* to get what I wanted from them, but I remember that it took me a very long time to *accept* the fact that I would just have to live with my frustrations. It took me back to the early days of 16-track, when the use of multiple microphones on drum kits was beginning to grow in popularity. Producers and engineers would get some very crisp and powerful tom-tom sounds, then go mad when they heard their wonderful sounds played back from the analogue 16-track machines. As with the sounds in the cathedral, it seemed like the better the sounds you began with, the more they were spoiled. We would, as a matter of course, call in the maintenance engineers to check the alignment, but they usually put a square wave into the machine and showed us on an oscilloscope display a total travesty emerging. We realised that the inherent phase shifts of analogue recording could forever limit us from hearing recordings of the sounds which we heard on the monitors, directly from the studio.

I was never quite sure whether it was something in the 'magic' sounds which the analogue recorders (or tape) could not handle, or whether it was the greater sense of loss at not being able to faithfully record the best sounds, that always seemed to me to be the reason why *those* sounds were the ones most noticeably destroyed. Then again, when digital recording should have solved this problem, no matter what was the cause, people began to complain that the digital recordings lacked the 'glue' that held the whole drum kit together, and often reverted to analogue machines for drum recordings. So was digital the magic word? I think not. It seems that every instance must be judged in the light of its own circumstances, and that absolute generalisations are difficult to make. Unfortunately, that over-used word 'compromise' rears its head all too frequently.

8.6 A compact studio

Possibly we should end this chapter with a look at a compact studio, where acoustic variability was needed, along with a good degree of separation. The studio was built in London, and was for commercial use. Its owner was a drummer, and, not too surprisingly, drum sounds had a strong input into the design considerations. Nevertheless, there was also great importance put on vocal recordings, and electric guitars and basses. Figure 8.12 shows the overall layout. The main studio area is subdivided by hinged panels which have hard surfaces on one side and soft, absorbent surfaces on the other. The whole studio is situated in a warehouse, with relatively few problems caused by a small amount of escaping sound. It is not on a main road, so the low frequency rumble of fast, heavy traffic is no problem either. A controlled leak of low frequencies was allowed to pass into the warehouse, thus preventing excessive build-up in the smallish rooms.

Drums can be recorded in the main area, in an acoustic ambience of whatever liveness or deadness that the room can provide with its variable states. Separation for guide vocals or other instruments can be achieved by creating booths out of the hinged partitions. Outside of the main recording area, and

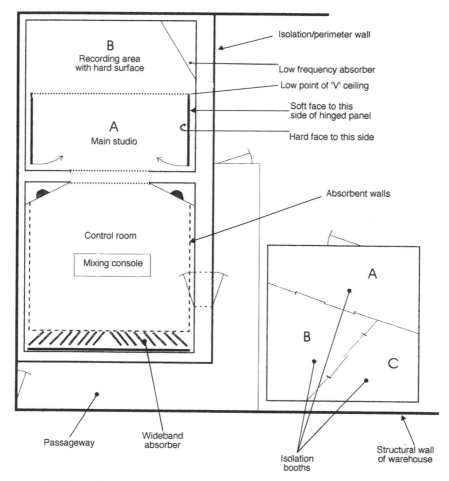

Figure 8.12 Studio in a warehouse, with external booths. With the hinged panels in the positions shown, the whole of area A/B becomes one, large live room. When the panels are swung out to the dotted line, section B remains live, but in section A, soft, absorbent surfaces become exposed, rendering the acoustic in that area much more dead. Panels can be moved individually, or partially, to change the sound or to use them as screens between instruments. The doors are quite heavy, with wheels supporting their outer ends, so isolation between sections A and B is quite reasonable when the doors are swung out to the dotted line. The ceiling is an inverted V-shape, with its apex above the dotted line. Over area A it is absorbent, and over area B it is finished with wooden panels.

The isolation booths, intended for high level recording of bass or guitars, are well isolated from the main studio and control room. Booths A and C are acoustically rather dead, whilst booth B is relatively live. Sliding glass doors are used to maximise the usable floor space. With windows in the control room entrance doors, and in the side wall of booth B, visual contact between the engineer and booths B and C could be provided.

If such a warehouse were not in a noise-sensitive location, the walls could be relatively lightweight, allowing a controlled leak of low frequencies, and thus giving the rooms better LF performance without the use of too much internal LF absorption. With a lighter weight construction, costs can also be greatly reduced

separated from it by a corridor, is a separately floated box, containing three small booths into which can be placed guitar or bass amplifiers, fed from instruments in the other rooms via line amplifiers. The musicians themselves can be located in the main studio, or the control room, which can also house keyboards and their attendant musicians. Once the rhythm tracks have been laid, the main room can be reconfigured for the recording of acoustic guitars,

Figure 8.13 Dobrolyot Studio, St Petersburg, Russia

Figure 8.14 The Pink Museum, Liverpool

(a) The form of a 'village' within the building

(b) The view from the stone room into the vocal room which can be fitted with curtains when necessary

lead vocals, backing vocals, or anything which may be required for completion of the recordings. This is yet another approach to the concept of a multi-room studio with a combination of acoustics.

8.7 Summary

Figures 8.1, 8.4 and 8.12 show three actual, workable, successful, but very different methods of tackling the problem of how to combine different acoustic options. The range of possibilities is almost endless, but these three examples give a feel for three very dissimilar approaches which could be variously adapted and modified to fit a wide range of likely starting conditions. Again, however, there are few absolute rules, which is why the experienced studio designers are so frequently relied upon for their expertise. Their function is not only to offer design suggestions, but also to point out things which the clients want, but which may lead to trouble in the future. Such points are not always obvious, and the problems are often only realised after past experience has pointed them out. Hopefully, this chapter will have at least begun to point out the strength of certain concepts, together with the weaknesses of others. As it has progressed, we have been able to look at some of the decisions which have had to be made along the way. One thing is for sure, however; the permutations are endless.

As a final point of possible interest, the rebuilt version of the aborted attempt to construct the studio, as depcited in Fig. 8.2, is shown in Fig. 8.13. The studio now incorporates a wooden live room, an acoustically dead vocal room, and a moderately large general studio area with reflective, absorbent and diffusive walls, not too dissimilar to the ideas outlined in Fig. 2.19. It works well, and the owners are very pleased with the results. It also ended up costing far less to build than the $50,000 which had already been spent on the originally proposed design.

The idea of a whole concept is illustrated in Figure 8.14. As shown in Fig. 8.14 (a) the studio takes the form of a 'village' within a building. The hexagonal room in the foreground is a vocal room which leads off a granite live room. The main neutral recording room is beyond, with skylights rising above it. At the back, with an external entrance on the extreme right of the photograph, is the control room. The balcony, from where the photograph was taken, has facilities for the preparation of food and drink, and also incorporates a lounge area with satellite television projected on to a large screen. The idea behind the design was to take the musicians into a self-contained space, isolated from the outside world, in which they can relax in absolute privacy. Despite the sense of spaciousness and tranquillity it is less than 30 m from a busy commercial street with shops and many restaurants of different ethnic origins.

References

1 Zenon Schoepe, 'Galaxy', *Studio Sound*, Vol 36, No 10 pp 42–44 (October 1994)
2 Earl Geddes, 'Analysis of the Low Frequency Sound Field in Non-Rectangular Enclosures using the Finite Element Method', PhD dissertation, Pennsylvannia State University, USA (1982)
3 Philip Newell, 'Studio Monitoring Design', Focal Press, (1995)

Chapter 9

The studio environment

9.1 Colour, light and human sensitivities

Much has been written in the preceding chapters about the studio rooms being, first and foremost, rooms for musicians to play in. Obviously, the individual tastes of each and every musician cannot be accommodated by any one studio design, so the 'ultimate' environment cannot be produced. Nonetheless there are a few general points which are worthy of consideration. Spaces finished in light colours, for example, feel larger than they would do if finished in dark colours. In general, I find that most musicians like light colours, as they tend to create less oppressive atmospheres in which to spend long periods of time. Daylight would seem to be another widely desired asset.

9.1.1 Daylight

When I was rebuilding The Manor studio in Oxfordshire, England, in 1975, an accidental collapse of part of the wall, above an old, small window, allowed sunlight to stream into the building. Eventually I decided to keep the windows open, but not before some deep consideration of the likely acceptance of daylight in the studio. It may seem strange now, but at the time, almost all studios were without windows. In 1971, The Manor had become one of only a very few serious recording studios in country locations. Perhaps the sound isolation requirements of the in-town studios had led, as a matter of course, to windows being bricked-up or otherwise sealed. In the original 1971 version of The Manor studio, Tom Newman had put a window in one side of the elevated control room. This took a few people by surprise, but it was soon generally considered to be a good thing to be able to gaze across the countryside from time to time. Remember this was a time when musicians were just beginning to have a little more say in how studios should be designed, and Newman was first and foremost a musician, rather than a studio designer.

I must admit that I had found it refreshing working in the control room of The Manor after years of being subjected to London basements. The awareness of changes from day to night, and more unpleasantly after very late sessions, from night to day, along with an awareness of the changing seasons, are all instrumental in maintaining a person's well-being. This fact is perhaps more medically recognised now than it was then. Once it had been established that the sound isolation in studios with outside windows was adequate, daylight soon became widely accepted as a desirable asset for studios. Since then, I have always fought to bring daylight into studios, if at all possible. In fact, over 90% of the studios which I have designed in the last ten years have had outside windows and natural light.

9.1.2 Artificial light

The use of lighting to create moods can be highly beneficial to the general ambience of a studio. Fluorescent lighting, apart from the 'hardness' of its light, is generally taboo in recording studios because of the problems of mechanical and electrically radiated noises. The mechanical noise problem can sometimes be overcome by the remote mounting of the ballast chokes (inductors) which prevent the tubes from drawing excess current once they have 'struck' (lit) but the problem of the radiated electrical noise can be a curse upon electric guitar players. In some large studios I have seen fluorescent lights used without apparent noise problems in very high ceilings (6 m), well above the instruments, but in general, they are likely to be more trouble than they are worth.

Lighting is a personal thing, of that there is no doubt, and the design of studio lighting systems is quite an art-form in itself. Personally, I must admit to not liking the low-voltage, halogen lights, as I find their light too stark, and they are not easy on my eyes. For me, there is still nothing quite like an argon filled, tungsten filament bulb. Figure 5.1 shows a room with floor-level, concealed, tungsten strip-lights, illuminating the stonework and casting shadows of the surface irregularities up the wall. Downlighting is provided by old theatre floods. Figure 5.2 shows the use of wall lights and an 'old' candelabra, to create the effect of an old castle. The room in Fig. 5.6 was expected to be used by many session musicians, so, in its ceiling, it had both tungsten 60° downlight reflectors and a skylight to allow in daylight. These forms of lighting would facilitate the reading of sheet music. On the other hand, for more mood, the skylight could be covered and daylight would then only enter via the stained glass window, which could also be externally lit at night to great effect. 'Mood' lighting was also provided by electric 'candles' on the walls.

In all of the above cases, the lighting could be controlled not only by the switches of individual groups of lights (or even single lights in some cases) but also by Variacs. Variacs are continuously variable transformers, which despite their bulk and expense (15 cm in diameter by 15 cm deep, and roughly US$60 each) are the ideal choice of controller in many instances because they cannot produce any of the problems of electrical interference noise which almost all electronic systems are likely to create from time to time. Variacs simply reduce the voltage to the bulbs, without the power wastage and heat generation of rheostats, which dim by resistive loss.

Electronic dimming systems using semiconductor switching can be a great noise-inducing nuisance, both via the mains power (AC) and by direct radiation. It is simply not possible to switch an AC voltage without causing voltage spikes. So-called 'zero voltage switching' may still not be zero current switching, as the loads which it controls will be unlikely to be purely resistive. Anyhow, a theoretically perfect sine wave can only exist from the dawn of time to eternity, and cannot be switched on or off without transient spikes. Even if you find this hard to believe, believe it. It *is* possible by more complex electronic means to produce a 'clean', variable AC, but the methods of doing so tend to be even larger and even more costly than Variacs, whose pure simplicity is a blessing in itself.

From the early 1970s to the mid-1980s I frequently installed coloured lighting, as well as white light for music reading and general maintenance. Suddenly, for a decade or so, it became *passé* but, recently, clients have again begun to ask me for coloured lighting. Such are the cycles of fashion. Anyhow, whatever lighting system is used, the general rule should be 'better too much than too little'. One can always reduce the level of lighting by switching or by voltage reduction; but if there is insufficient total lighting for music reading or maintenance, work can become very tiresome.

9.1.3 Ease and comfort

General comfort is also an important issue, as again, comfortable musicians are inclined to play better than uncomfortable ones, and few things seem to kill the sense of comfort and relaxation more than an untidy mess of cables and other 'technical' equipment. Musicians should never be made to feel in any way subservient to the technology of the recording process. The studio is there to record what *they* do; they are not there to make sounds for the studio to record. Sufficient sensitivity about this subject is all too frequently lacking in studio staff, especially in many of those having more of a technical background than a musical one.

Easy access to the studio, for the musicians, is another point worth considering well when choosing the location for a studio. I do know of studios which have been on the fourth floor of buildings with no lifts, but their existences have usually been rather short-lived. Especially for musicians with busy schedules, easy access and loading, together with easy local parking, can be the difference between a session being booked in a certain studio, or not. It is also useful to have some facility for the storage of flight cases and instrument cases *outside* of the recording room. Apart from the problems of unwanted rattles and vibrations which the cases may exhibit in response to the music, when a studio looks like a cross between a warehouse and a junk room, it is hardly conducive to creating an appropriately 'artistic' environment in which to play.

The points made in this section are not trivial, though they are not given their due attention in all too many designs. Experienced studio owners, operators, engineers, users and designers appreciate these points well, but in the current state of the industry, a large proportion of studios are built for first-time owners. As so many of them fail to realise the true importance of these things, they are frequently the first things to be trimmed from the budget, especially when some new expensive electronic processor appears on the market midway through the studio construction. Most long-established studios do provide most of these things though, which is probably one of the reasons why they have stayed in business long enough to *become* long-established.

9.2 Ventilation and air conditioning

It is certainly not only in studios that people complain about the unnatural, and at times uncomfortable, sensation of air-conditioned rooms. Unfortunately though, with sound isolation usually being good thermal isolation, and given that people, lights and electrical equipment produce considerable amounts of heat, air conditioning of some sort is more or less a mandatory requirement in all studios. In smaller rooms, such as vocal rooms, where few people are likely to be playing at one time, sometimes a simple ventilation system is all that is required. In fact, I have found it advantageous in many instances to provide a separate, well-filtered, ventilation-only system, in addition to any air-conditioning. It is remarkable how many times these systems have refreshed the musicians, without having them complain of dry throats.

9.2.1 Ventilation

In order to get the best out of a ventilation system there are a few points worth bearing in mind which are of great importance. One of the most important rules is never to only extract air from the room. If an extraction-only system is used, the room is in a state of under-pressure; a partial vacuum. Whenever a door is opened, or for that matter, via any other available route, such as via air-conditioning systems, air will be drawn into the room. This air will be dirty, as it cannot be filtered, and dust, dirt, and general pollution will enter the room along with the air, playing havoc with singers' throats, and leaving dust everywhere. The air conditioning *inlets* will almost certainly be filtered, but in almost all cases, the outlets will not be, so air which a ventilation system pulls in through an air-conditioning *outlet* may be dirty. It may even pick up more dirt from the outlet duct itself, which may be full of dust and tobacco deposits.

In rooms with only a low volume of air flow, such as vocal rooms, it is often possible to simply use an input-only fan, with the air being allowed to find its own way out via an outlet duct of suitable dimensions. When high volumes of air-flow are necessary, though, extraction fans are usually fitted, but there are some basic rules which should be followed. For the reasons mentioned in the previous paragraph, the extraction fans should never be operating without the inlet fans, as they would constitute an extract-only system. What is more, the flow rate of the extraction system should never equal or exceed that of the inlet system, as this would still produce an under-pressure in the room. It is prudent to restrict the flow rate of the extraction system to between 60% and 80% of the inlet flow, depending largely upon the degree to which air can find its way out of the rooms by routes other than the intended outlet system. In purpose-designed studios, the extraction flow rate can usually be quite high, as the rooms are normally more or less hermetically sealed at all points other than via the air-change system(s). However, multi-functional rooms, which may serve as studios from time to time, are often somewhat more leaky through their doors, windows, and roofs in particular. In these cases, a proportionally lower extract flow rate would be desirable, as over-pressures will be more difficult to achieve.

The normal way to ventilate is to draw air in from outside through a filtration system. This would usually consist of either a single filter, or a series of filters, with removable filter elements which can easily be cleaned and/or replaced. Air will then pass to an in-line fan and on to a silencer, or series of silencers, before entering the room. This way, the room is kept in over-pressure, and any time a door is opened, the clean, filtered air in the room will escape through the door, keeping the dirty, outside air from entering through the doorway. Such systems keep the rooms cleaner, and ensure that the air which enters the rooms is always clean. Outlets will normally pass first through silencers (both to prevent street noise from entering, and studio noise from leaving), continuing on to the outside air via a back-draught damper. The damper is a one-way flap arrangement, which only allows air to flow out. If the ventilation system is switched off, and the wind is in an unfavourable direction, then should air attempt to enter via the outlet, the flaps close off the ducts. This also prevents dirt from entering via the unfiltered outlets. Fire dampers are sometimes also installed, which close the air ducts, completely, should the room temperature rise above a pre-determined limit. By this means, if the temperature rise is due to a fire, the oxygen supply to the fire will be cut off, thus slowing the spread of the fire even if not extinguishing it completely. Another point to remember is to turn off the ventilation system when the studio is unoccupied so that if a fire should begin, it will not be supported by a constant supply of oxygen.

The ventilation ducts will often be of the 'acoustic' type, having a thin, perforated aluminium foil liner, then a wrapping of 5 cm glass fibre, and finally an outer aluminium foil covering. In this type of ducting, the air is allowed to pass through a relatively smooth inner tube, which presents only minimal friction, and hence loss of air flow. This smooth inner lining must be as acoustically transparent as possible to allow the sound to be absorbed by the fibrous layers around it. If the sound cannot easily penetrate the inner lining of the duct, it will travel very effectively along the tube. Sound can travel along kilometres of smooth bore tubing with remarkably little loss, as no expansion of the wave can take place. This is the principle by which the speaking tubes of the old ocean liners were able to provide excellent communications, often over long distances and in noisy surroundings; bridge to engine-room, for example.

Duct *size* is also very important, as for any given quantity of air to pass, the air flow down a duct of large diameter will be much slower than down a narrower duct. A 20 cm diameter duct has an approximate cross-sectional area of 315 cm^2, and a 30 cm diameter duct an area of 709 cm^2, which is more than double. Therefore, for any given rate of air flow, the speed of the air down the 30 cm duct need only be just under half of that down the 20 cm duct. As the noise caused by turbulent air flow follows something like a sixth-power law, then for any given flow rate, ventilation system noise will rise rapidly with falling duct diameters. For adequate air flow, even in the smaller rooms, 20 cm seems to be about the minimum usable duct diameter.

9.2.2 'Split' air-conditioning systems

Traditionally, professional recording studios have used conventional, ducted air-conditioning systems, and noise floors of NC_{20} or less have been achievable, even with high rates of air flow. Such systems are still the only way to properly air condition a studio, but as has always been the case, they are relatively expensive. Since the early 1980s, the real cost of studio equipment has been falling, and somewhat unrealistically, the charges per hour for the studios have plummeted. At a time when a million pounds is spent on the recording equipment for a studio, to pay £100,000 for an air-conditioning system may not seem too disproportionate. On the other hand, when manufacturing technology is such that it is possible to buy for £100,000 the equipment that will not perform far short of the £1m equipment, people seem to balk at the charges for good air conditioning. Competition between studios has forced real prices ever downwards, and a state of affairs has now developed where the cost of ducted air-conditioning systems, for most of the mid-priced studios, has become insupportable.

Largely for economic reasons, there has been a great increase in the number of 'split' type air-conditioning systems coming into studio use. Although these are by no means ideal for this purpose, they are many times cheaper than the ducted systems, but as the heat exchangers and their fans are in the studio, with only the compressors remaining outside, there is an attendant noise problem. In control rooms the units can usually be left running in 'quiet' mode, as the noise which they produce is often less than the disc drives and machine fans which may also be in the room, but in the studio rooms, they must be turned off during most recordings. Unfortunately, this intermittent use can lead to temperature fluctuations, which may not be too good for the consistency of the tuning of the instruments. Nevertheless, despite their problems, there are now many split systems in studio use.

Studio customers have become used to inexpensive recording, and all too few of them now want to pay for facilities which can provide all the right conditions for optimal recordings. Unfortunately, as this has become widespread, it has also apparently become *acceptable* in these 'market forces' days. Air-conditioning systems have, in very many cases, been tailored to a 'reasonable' proportion of the recording equipment budget, which has led to unsatisfactory air conditioning, but this reality exists.

There are limits, however, below which the lack of suitable environmental control will pose serious problems, not only for the musicians but also for instruments such as pianos and drums. Draughts of air are generally disliked by musicians, although different musicians, or groups of musicians, may have some differences in their preferences of optimal temperature for their comfort. If a studio is block-booked for a week or so, then the chosen temperature can be selected at the start of the session, and maintained, but pianos should not be tuned until the temperature has had time to stabilise. With shorter bookings though, the temperature should be kept at a suitable compromise, as frequent changes in air temperature will wreak havoc with the tuning of any permanently situated pianos, and many other instruments for that matter.

Humidity is also another factor which needs consideration. If maintained too low, it can dry out the throats of singers, and may cause piano sound boards to crack. If it is too high, it can be uncomfortable for the musicians, and corrosive to instruments. A humidity level of 70% is good for most purposes, and in the better studios, regular attention is paid to the maintenance of the appropriate levels. Unfortunately, in an enormous number of smaller, or less professionally operated studios, little or no attention whatsoever is paid to humidity control. Once again, the relentless driving down of studio prices has rendered it impossible for many 'professional' studios to provide the sort of environmental controls that are musts for *truly* professional studio operations.

There is now a huge 'consumer' recording market which, although providing standards below those to which the 'professional' market was accustomed, has blurred with the professional market to such an extent that its lower standards have begun to influence the professional world. Some of this has no doubt been due to the enormous influence of electronically based music, for which studio acoustics and air-conditioning noise have not been problems, yet I fail to understand how some people can hear any detail at all in control rooms where the noise floor is made ridiculously high by the presence of hard-disc drives and numerous equipment fans. Fortunately though – at least *I* believe it to be fortunate – there seems to be a swing back to the use of acoustic and electrified (as opposed to electronic) instruments, which is just as well if much of the passed-down experience that exists in the recording industry is not to be forgotten through the loss of a whole generation of recording staff to the computer world.

9.3 Equalisation and foldback

The great emphasis which has recently been placed on 'synthetic' instruments has totally bypassed the studio acoustics and microphone-positioning side of the recording art. With computer-generated sounds, there are two ways of altering them, either inside the generation systems themselves, or via equalisation. Despite whatever lip service may be paid to the niceties of microphone positioning, I personally feel that, in too many cases, too much reliance is currently put on equalisation in the search for appropriate sounds. There seems to have always been a tendency for recording personnel in Europe to rely more on equalisation than their counterparts in the USA, or, as I am now finding out, perhaps in Russia also. Equalisation is a useful tool to have, but if the microphone does not collect a good sound in the first instance, then this limits what can be achieved by any further equalisation.

9.3.1 Equalisation considerations

The term 'equalisation' originated in the film industry. If close-up shots were recorded with the microphones reasonably close, and then a subsequent shot, still relatively close, required the microphone to be moved away, out of view of the camera, the sound would change, and the editing of the shots would sound unnatural. Tone control devices were then used to adjust the tonal balance of the recordings from the two different microphone positions, to 'equalise' their apparent distances with relation to the picture. In general, it was much later in the USA than in Europe that equalisation on all channels became the accepted norm on music-mixing consoles. I recall working in one of the studios of CBS in New York in 1976 with a mixing console which had no on-board equalisation whatsoever; it all had to be patched in when required.

I think that in Europe there was a certain laziness which developed from having equalisation generally available on all channels from a much earlier date. It became interesting to experiment with this new technology, and it certainly became easier to turn a knob, rather than to go to the studio to change a microphone or its position. On the other hand, the different musical development at the time could also have played its part. Certain things can develop at critical times, and their influence can be great or small, depending on exactly when and where they take place. It is also possible that the much larger film industry in the USA provided a greater pool of knowledge about recording techniques. To this day, though, I personally feel that equalisation is a last resort, though so many recordings have now become so stylised by their equalised sound that equalisation has become inextricably linked with much of the recording process. This is fine, as long as the over-reliance on it does not lead to an almost complete loss of knowledge of how to record *without* equalisation.

9.3.2 Headphone foldback

The topic of equalisation may seem to be in a strange place in a chapter about the studio environment, but there is part of the studio environment that passes through the mixing console in the control room. It is an extremely important part of the recording process, because it can be the entire acoustic reality for the musicians: the foldback. In by far the majority of cases, musicians will record whilst wearing headphones. In these instances, the acoustics of the studios can only be heard via the microphones, mixing console and headphones, so the musicians can find themselves in a totally alien world if due consideration is not given by the recording engineers to the creation of the right foldback ambience.

If musicians need to hear in the studio, then they need to hear it in their headphones as well. Many musicians play off their own tone, so if they cannot hear themselves as they need to, they may 'force' their tone or perhaps hold back, and neither is satisfactory for optimal recording. If musicians need to hear richness in the sound, then they ideally would need to hear it in their headphones; if they need the reinforcement of lateral reflexions, then they need to hear those in the headphones also. At times, it can be considered beneficial to put up ambience microphones which will be used in the foldback only. If this helps to improve the sense of 'being there' for the musicians then it is surely worthwhile, though it is rarely done.

There are two big problems with the optimisation of foldback. Firstly, few engineers have spent enough time as recording musicians to be sufficiently aware of the complexities and importance of the requirements of different musicians in terms of foldback. They cannot be blamed for this, as they cannot spend their lifetimes doing two things at once. The reverse of course applies; so many musicians are used to poor foldback that they fail to realise what *could* be achieved. Neither the recording engineers nor the musicians have spent a working lifetime doing the other end of the foldback job. There are times when musicians are given their own foldback mixer for use in the studio, but they rarely have access to the reverberations that are available from the control room and which can greatly aid their perceptions of space.

Secondly, however, if too much of the foldback burden is loaded on to the musicians, then they can become distracted from their primary job: playing. In fact, one significant restriction on foldback balance is time. The foldback mix cannot be set up until the musicians are playing, but once they *are* playing, too much time cannot be spent setting up the foldback before they 'go off the boil' and lose their motivation to play. Furthermore, too much fiddling with the foldback, with levels going up and down and things switching in and out, is absolutely infuriating for the musicians. Remember, effectively it is their whole audible environment that is being disturbed. For them, it is like a painter trying to paint a picture with the lights randomly going on and off.

Where possible, foldback should be very carefully considered, and a stereo foldback system is much easier to hear clearly than a mono one. In stereo, even things which are perhaps a little too low for the ideal balance for a mono mix can be perceived much more easily due to the spacial separation which stereo provides. Whenever possible, I like systems where the recording engineer can monitor exactly what the musicians are hearing, which includes listening at the same level. Obviously, on systems where all, or at least many, of the musicians are in a position to make their own balances, this is hardly practicable, but in cases where many headphones are driven from a common power amplifier, it is. In these cases, it can be beneficial for the engineer to have a line in the control room which is connected to the same power amplifier output as the studio headphones, and, where possible, to monitor on the same type of headphones that are being used by the musicians. There can, in this way, be little doubt that the engineer is monitoring the selfsame space that the musicians are immersed in, and there is therefore much less chance of misunderstandings.

In many cases where the control room foldback monitoring system is via headphones plugged into the mixing console, or where it is heard on loudspeakers, I have frequently witnessed complaints from the musicians about the foldback being unsatisfactory (and indeed it has been so), yet the control room foldback monitoring has not revealed the problems. This has often led to time wasting, or, if the problems have not been pursued, to the musicians having to try to play with the problems unresolved. In either case, the musical performance will probably suffer. The musicians also seem to feel an added sense of being understood and appreciated when they feel that their environment is being shared by its controller, and this helps to allay any insecurities which they all too frequently feel.

The 'virtual' spaces in which the musicians may have to perform can be just as important to the recording process as the very real spaces in which they physically play. A full understanding of this is a fundamental requirement of being a good recording engineer, and good foldback systems are an equally fundamental part of a good studio. In effect, there are therefore three aspects to the acoustics of a recording space: firstly, the acoustic as heard in the room; secondly, the acoustic as collected by the microphones; and thirdly, the combination of the two as perceived by the musicians if they must use headphones. All should be considered, very carefully, in both the design and the use of the rooms.

Choice of open or closed headphones can also modify this environmental balance, as closed (sealed) headphones tend to add to the feeling of isolation, and take the musicians one step further away from their 'real' acoustic space. Nonetheless, there are times when closed headphones are very necessary. Drummers, for example, may need closed headphones in order to avoid having to use the painfully high foldback levels which may be necessary so that the other instruments can be heard over the acoustic sound of the drums. In this case, closed headphones are used to keep unwanted sounds out. Conversely, vocalists, or the players of quiet, acoustic instruments, and in particular those instruments which require the positioning of a microphone close to the head of the musician, may also need to use closed headphones. This is especially the case during overdubbing, when the 'tizz tizz', reminiscent of sitting next to somebody using a personal stereo system, may be picked up by the microphone, and may subsequently be difficult to remove from an exposed track. In this instance, the closed headphones are used to keep the unwanted sound *in*.

It is a pity that foldback is so often subject to so much compromise, but in a way, such is its nature. As a great deal of modern music has developed out of past recording technology, it is not surprising that aspects of the limitations of the older technology should be carried along. But, it is imperative to remember that whatever wonders may be created in the acoustic design of the studio, often for the benefit of the musicians, as soon as they put on a pair of headphones, those muscians can be in a different world, and it is best not to leave them feeling lost in it. This is one very important difference between designing live performance spaces and recording spaces. In the latter case, it is not usually how the musicians hear the space which predominates the design considerations, but how *microphones* react to the space. This point should never be forgotten, but all too frequently it is.

9.3.3 Loudspeaker foldback

Many studios and musicians find the facility of being able to provide foldback via a loudspeaker a useful asset. The concept of the 'tracking loudspeaker' goes back further than my own involvement in the recording world (1966) and I remember Lockwood cabinets, dedicated solely for that purpose, being used at Pye Studios in London, in 1970. Many vocalists found that it was easier for them to perform without headphones so that they could clearly hear their own, natural voices. A loudspeaker would be positioned facing the vocalist, and the backing track would be played back through the loudspeaker at the minimum level needed to allow the singer to perform well, but not so loud that the overspill into the microphone would cause problems. The directional characteristics of a cardioid microphone were employed, in order to reject as much of the loudspeaker output as possible. This was further aided by placing non-reflective screens behind the vocalists, to prevent the playback signal from bouncing off a wall behind and subsequently entering the microphone.

These loudspeakers also became useful as a means of playing a recording back to the musicians without them having to leave their positions and go into the control room, which would in any case perhaps be too small to house them all. This was never a particularly good means of assessing the *sound* of the recording, but was a useful tool for discussing performance quality or mistakes with the producer. Loudspeakers were also useful in orchestral recordings when the conductor needed to highlight some point in the performance to the musicians, as taking a whole symphony orchestra into the control room would clearly be out of the question. Once again, ease and comfort go a very long way towards getting the best performances out of musicians, so they should be given great attention when designing *or* operating a studio. The recording process begins with the musicians, and they are the foundation on which the rest of the process is built. If these foundations are weak, the entire edifice will never achieve its potential strength or quality. Always remember that!

Chapter 10

Limitations to design predictions

10.1 Room responses

Much of the space in this book could have been filled up with corresponding plots of the reverberation times (RT_{60}) and the energy/time curves for each of the rooms discussed and for each of the photographs shown. The waterfall plots for each room could also have been discussed, and indeed, many people would perhaps have expected such, but specifications often tell us very little about the perceived sound characteristics that we have been discussing. What is more, in the wrong hands, they can be misleading, so perhaps we had better take a look at some of the different representations, and their different uses.

The classic RT_{60} is the time taken for the sound of an event in a room to decay to one millionth of its initial power, which is -60 dB. The typical RT_{60} (reverberation time) presentation is shown in Figs 10.1(a) and (b). The two plots shown are actually the reverberation time responses for two famous concert halls within 20 km of central London. In these plots, reverberation time is plotted against frequency, and these curves are useful to see the 'frequency response' of the halls. It is not surprising that the subjective quality of (a) is warmer and richer, but less detailed than (b), as (a) has a much longer low frequency reverberation time, which not only serves to enrich the bass, but can also mask much low level transient and high frequency detail. Unfortunately, such plots only tell us what is happening at the -60 dB level, and not what happens during the decay process itself. Depending on *how* the reverberation decays, it is possible that seemingly obvious deductions of the subjective quality could have been erroneous if based only on RT_{60} information.

Besides the classic representations such as those in Fig. 10.1, which plot time against frequency, there are other ways of looking at the reverberation decay, such as methods which plot reverberant *energy* against time. One such representation is the Schroeder plot. Figure 10.2(a) shows the decay 'curve', or Schroeder plot, that would be expected from a good reverberation chamber. However, in studio rooms there is almost always absorption, diffusion, and a whole series of reflexions, both of the early and late variety, which all serve to modify the smooth decay of a true reverberation response.

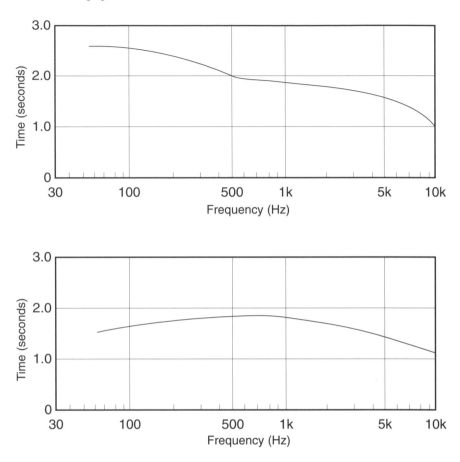

Figure 10.1 Reverberation times of two famous concert halls in England

Plots more typical of studio rooms are shown in Fig. 10.2(b). Figure 10.3 shows a series of 'reverberation' responses from different rooms, all of which have nominally the same RT_{60}, and all of which could conceivably produce very similar plots to each other, of the type shown in Fig. 10.1. Clearly, the solid curve contains less overall energy than the others. In a room showing that type of characteristic response, with an *initial* decay that is much more rapid, there will be less of a tendency for the reverberation to mask the mid-level detail in any sounds occurring in the following half second after the occurrence of a loud sound. The broken curves, on the other hand, would represent spaces which sounded richer than the one represented by the solid line. In many ways, it is the *early* decay time (the time taken for the sound to decay to 10 dB below its initial level), which tells us much more about a room than its RT_{60}. (Incidentally, do not confuse the early decay time with RT_{10}; the latter is the -60 dB level, extrapolated by drawing a continuation line from zero, through the -10 dB point and down to the -60 dB level. It is used in cases where background noise problems preclude the actual measurement of the RT_{60}.)

In small rooms, the truly diffuse sound-field required to produce reverber-

Figure 10.2 Schroeder plots

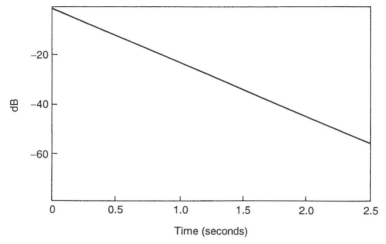

(a) In a perfectly reverberant room, the Schroeder plot would show a straight line decay. In the case depicted here, the RT_{60} is slightly over 2.5 seconds

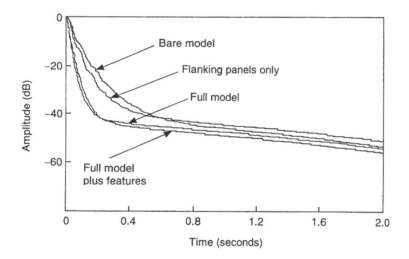

(b) This Schroeder plot of a test room shows how the installation of acoustic control items removes the energy from the early part of the decay curve, 'cleaning up' the room, but without significantly reducing the decay time at the −50 dB level

ation can never develop, and relative energy in the modes, reflexions and diffused sound all contribute to the overall perceived sound. The Schroeder plots of Figs 10.2 and 10.3 show the overall envelope of the decaying energy, which is a very useful guide to the general behaviour of a room, but when problems exist, it is sometimes necessary to see more detail. Figure 10.4 shows an energy/time curve (ETC), in which the individual predominant reflexions can be seen as spikes, above the general curve of the overall envelope. These plots, like the Schroeder plots, show level against time, so the

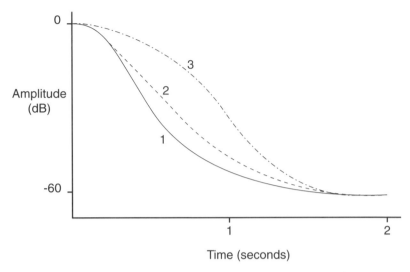

Figure 10.3 Three different decay characteristics, all with nominally similar T_{60} values of 2 seconds. Clearly curve 3 contains the greatest total energy, so if these curves represented rooms, then room 3 would sound 'louder', but room 1 would be likely to allow greater perception of detail

time between the initial event and any problem reflexions can be determined. When the time of travel is known, the path from the sound source to the microphone can be calculated. Any offending surface can then be located, and if necessary, dealt with.

10.1.1 The envelope of the impulse response, and reverberation time

Schroeder plots and ETCs are both representations of energy versus time, but they each have their distinct uses. They are also generated in different

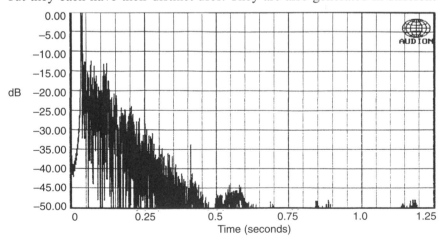

Figure 10.4 Energy/time curve (ETC). This ETC is a representation of a room not too dissimilar from the bare model plot in Fig. 10.2(b). The bumps at 0.55 secs, 0.85 secs and 1.15 secs were due to traffic noise. The plot was from an unfinished room.

Figure 10.5 Impulse response of room (bare model) shown in Fig. 10.2(b)

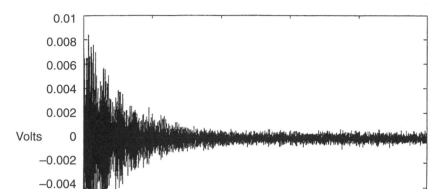

(a) The impulse response shown here does not appear on first glance to relate directly to the more intuitively apparent presentation of the Shroeder plot of Fig. 10.2(b), though both were generated from the selfsame measurement. The asymmetry of the upper and lower halves of the plot can clearly be seen, but due to the long time-window, the zero crossings are not obvious

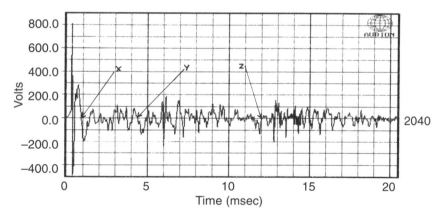

(b) When the time scale of an impulse response is stretched (in this case, the time scale is only 20 msec) the multiple crossing of the zero amplitude line can clearly be seen, such as at points X, Y and Z, and many other points

ways, in order to best highlight different aspects of the responses. Figure 10.5 shows the impulse response of a room – a simple representation of sound pressure against time. It may at first seem plausible that one half of this plot could be 'smoothed' to give a representation of the decay of the room, but it will be noticed that the plot is not symmetrical about its horizontal axis. It will also be noticed that there are many crossings of the horizontal axis, where the respresentation of the sound pressure level is therefore zero. If we simply summed the two halves, or overlaid the lower half of the plot on top of the upper half, then the zero-crossings, such as are indicated at points X, Y and Z on Fig. 10.5, would still retain their zero values. On listening to the decay of a room, it soon becomes intuitively obvious that the energy decay does not alternate between rapid bursts of energy separated by points of zero energy, but that energy is present in the room from the inception of the sound until its decay to below audibility. For this reason, some integration is necessary if the simple instantaneous sound pressure representation of the impulse response is to be made into a better representation of energy decay.

A measure of the power, or energy per unit time, in such a signal can be calculated by squaring the signal and then averaging it over a suitable length of time to yield the mean-square value of the signal (the familiar rms [root-mean-square] value is the square-root of this). The mean-square is a continuous positive value, which is only zero if the signal is zero for longer than the averaging time, and thus it does not contain the multiple zero crossings present in the original signal. Provided the averaging time is long enough, *continuous signals*, such as a sine wave, have a mean-square value which is independent of time and independent of the length of the averaging time. However, estimates of the changing mean-square value of a *transient* signal, such as an impulse response, can be *very* dependent upon the length of the averaging time.

10.1.2 Schroeder plots

In the early 1960s, Manfred Schroeder[1] had been frustrated in his attempts to accurately measure the reverberation time of some large concert halls. He was having trouble achieving repeatability in his results because, depending on the precise phase relationship of the elements within the random excitation noise, and the beating between the different modal frequencies thus excited, he could get some very different readings which could even halve, or double, on subsequent measurements, so confidence in the measurements was not good. The method which he developed to overcome these problems involved the excitation of the room by a filtered tone burst, and the recording of the signal onto a magnetic tape recorder. The tape was then replayed backwards, and the output was squared and integrated by means of a resistor/capacitor integrating network. The voltage on the capacitor would then represent (in reversed time) the averaged energy decay of the room. Because the squared tone burst response is a positive function of time, its integral is a monotonically decreasing function of time. The resultant plot is therefore a gradually falling line, free of the up-and-down irregularities of many other decay-time representations, and is thus easier to use for accurate decay-time *rate* assessments.

Nowadays, the generation of Schroeder plots is almost always by means of digital computers, which have made redundant the need for cumbersome tape recorders in the making of these measurements. The use of Schroeder plots is widespread because they show, perhaps more clearly than any other decay-time representation, the presence of multiple slopes within a complex decay tail. It is now widely appreciated that the initial slope, for the first

10 dB of decay, is perhaps one of the most important characteristics in the subjective assessment of the performance of a room.

The decay curve for a Schroeder plot is derived by squaring the decaying signal, and integrating backwards in time from that point when the response exceeds the noise floor (t_1), back to the start of the response (t_0). The value of the decay curve at any point in time (t) is therefore the integral of the response over the interval from t to t_1. The resultant decay curve is a smoothly varying, and (necessarily) monotonically decreasing function of time. This characteristic is a necessary condition of the total sound energy in a system (such as a room) after the source of energy has ceased. The technique is thus valuable in determining the decay slopes, particularly in the early part of the response, which are necessary for reverberation time estimates. Any fine detail in the impulse response is lost, however, due to the time integration.

For anybody unfamiliar with the term 'monotonic', it means either forever rising or forever falling, although fluctuations in the *rate* of rise or fall can be allowed. The monotonically falling characteristic of the Schroeder plot decay curve is easy to appreciate, as once the energy source has been turned off, the absorption in the room can only lead to a gradually reducing net energy level in the decay tail.

10.1.3 Energy/time curves (ETCs)

The energy/time curve is the result of applying a technique to a transient signal which does *not* depend upon time averaging, but which yields an estimate of a mean-square-like value which varies with time. The seminal paper on this technique was written by the late Richard Heyser, and was published in 1971[2]. As mentioned previously, the multiple zero-crossings of the impulse response do not represent points of zero energy, so before we look in more detail at the generation of an ETC, perhaps we had first better think about why an instantaneously zero SPL is not necessarily representative of zero energy.

Let us consider a simple pendulum. When the pendulum is at either end of its swing, it is at its maximum height but has zero velocity. When the pendulum is at the bottom of its swing, it has minimum height but maximum velocity. The two forms of energy present in the pendulum are potential energy, which is a function of the height of the pendulum above equilibrium, and kinetic energy, which is a function of the velocity of the pendulum. As the pendulum swings, there is an alternating transfer of energy from potential to kinetic to potential, with one form of energy being a maximum when the other is a minimum. The total energy in the pendulum at any time is the sum of the potential and kinetic energies, and is independent of time unless the pendulum slows down. A graph of either the height of the pendulum, or its velocity, would show multiple zero-crossings and would not therefore be a measure of the total energy in the pendulum. One could, however, obtain a good estimate of the total energy by estimating one graph from the other; calculating the two energies and then summing them together. This is the method used in the calculation of the ETC.

A signal, such as the impulse response of a room, is treated as though it represents one form of energy in an oscillatory system such as the pendulum above. Using powerful signal processing techniques, such as the Hilbert Transform, it is possible to derive a second signal which represents the other

Figure 10.6 Waterfall plots

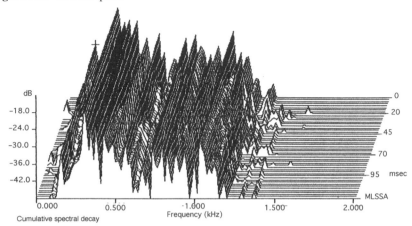

(a) This waterfall plot is of the bare model shown in Fig. 10.2(b). The cumulative spectral decay is represented in a three-axis format, the vertical axis representing amplitude, the front/back axis representing time, and the left/right axis representing frequency. In the case shown, only the low frequency range was being studied

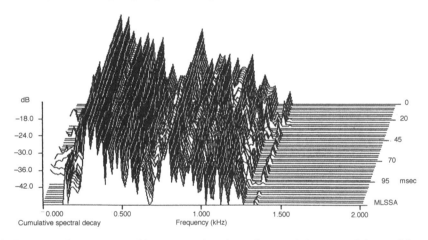

(b) For comparison purposes, this representation shows the cumulative spectral decay of the full model shown in Fig. 10.2(b)

form of energy. The two signals are similar, but 'appear' to be 90° out of phase with each other; one has a maximum (or minimum) when the other has a zero crossing, and vice versa. The ETC results from squaring these two signals and adding them together.

The ETC does not rely upon time averaging (or integration) and therefore does not mask any fine detail in the instantaneous energy of the signal. In general, the ETC of an impulse response has a characteristic which can increase as well as decrease with increasing time (it is *not* monotonic), and is thus useful for identifying early reflections and echoes.

To summarise, the Schroeder plot is most suitable for reverberation-time estimation where the main point of interest is the rate of decay of the energy in a room. The ETC is better for identifying distinct reflections, or other time-related details of a specific impulse response, such as those generated by a loudspeaker system.

10.1.4 Waterfall plots

Some of the characteristics of the ETCs *and* the classic RT_{60} plots can be combined into what is commonly known as the waterfall plot. Normally computer-generated, these plots show a perspective view in a three-axis form, as shown in Fig. 10.6. The vertical axis represents the amplitude of the sound, and the two horizontal axes represent time and frequency. Such plots are very useful, but care should be taken when assessing them. On first impressions, it often seems that they contain all the values represented by the three axes, but the nature of their perspective representation can hide some details in 'valleys' which lie behind some of the 'hills'. Nevertheless, waterfall plots are very useful forms of analysis, as in a single plot they allow an 'at a glance' assessment of reverberation time against frequency, decay rates at different levels *and* frequencies, and not only the ability to see discrete reflexions, but also the ability to assess the predominant frequency bands inherent within any individual reflexions. Despite all of this information though, one must always remember that what a room *sounds* like can still not accurately be assessed from *any* piece of paper, as measurement microphones, no matter how sophisticated, are not ears, and neither are they connected to human brains.

10.1.5 Directional effects

Unfortunately, even when supplied with the RT_{60} information about a room, *and* the Schroeder plots, *and* the ETC details, *and* the waterfall plots, this still tells us nothing about the diffusion or diffraction effects which may be taking place within the curves. We would still know little about any directional effects, which audibly may make one seemingly objectively innocuous reflexion more subjectively objectionable than an obviously higher level reflexion approaching the microphone position (or listening position) from a direction which causes only little offence to the ear. Floor reflexions, for example, can be collected by a measuring microphone at a high level, yet our sensitivity to vertical reflexions, in general, is far less than to horizontal ones of the same intensity. Human beings have poor vertical sound discrimination, as our vertical position is usually known, and evolution has caused the development of horizontal discrimination at a more urgent pace. The knowledge of the proximity and direction of predators or prey has been a great survival advantage, and because humans, their predators, *and* their prey have historically largely been confined to living on the ground, two-dimensional localisation has been all that has been needed.

Another directional effect which is *very* difficult to demonstrate in overall response plots is the effect of any coupling between acoustic modes and structural modes. Any surfaces immersed in an acoustic energy field will, to some degree or other, vibrate as a result of the coupling to the acoustic energy. Sections of the structure of a room, or panels attached to the surface of the room, may vibrate and re-radiate energy into the room.

The re-radiation will not only modify the modal interference (standing wave) patterns of the room, but will also act as a secondary sound source. To a person in the room its effect may well be audible, and detectable from a definite direction. This is another subjectively undesirable possibility which would be difficult to locate in most visual representations of room responses. It will no doubt by now be realised that if we wish to find out how a proposed room may sound, in advance of its construction, simply building to some preconceived specifications will obviously, in the light of the previous paragraphs, not suffice. If we need to know how a room may sound, we need to resort to other techniques.

10.2 Scale models

Short of building full-size prototype rooms, the next best possibility would seem to be to build and test scale models. At concert hall level, this is a proven method, at least as far as the major characteristics of the room are concerned. In a one-tenth scale model of a studio room, music can be played through miniature loudspeakers in the model room, speeded up by ten times, and recorded via miniature microphones placed in the ears of a one-tenth scale head. The sounds can be recorded, and then slowed down by ten times, and the result, heard on headphones, should be a reasonable representation of what the full-scale room would sound like. The above description of the technique is something of a simplification of the whole process, but the technique is useful, and is used on some large-scale projects. One-fiftieth scale modelling is also used for larger halls, but because the frequencies must be scaled up by the same amount, air absorption losses are such that nitrogen is usually used to fill the models, and in many cases, predictions are restricted to the low-frequency range of the final, full size room. However, with rooms of studio size, the scale modelling methods would be expensive to put into operation. Furthermore, as the characteristics of acoustically small rooms are so dominated by surface features, which will not easily scale, the scale modelling technique is likely to fail. Whilst frequencies, sizes and general modal responses *can* be scaled, the effects of absorbent treatments, the irregular surfaces of stonework or the effects of floor resonances cannot. Mineral wool is almost impossible to scale, as are carpets or curtains. Scale models of small rooms can only be used to determine gross effects, and if effects are gross, they can normally be deduced from pure theory.

10.3 Computer models

The power of computer modelling seems to increase day by day, but computers still cannot *design* rooms. They can be an aid to the experienced designer, who knows what he or she is looking for, but there is always a danger that in less experienced hands, the graphics and apparent simplicity of the operation of the programmes can lead to a belief that they give all the answers. They do not; the underlying rules and calculations behind the programmes must be understood if gross over-simplification and over-confidence in the results are to be avoided. In many cases, exact solutions are only possible for simple shapes, whereas in practice the same things which cause problems in the scaling of physical models will cause similar problems with computer modelling. Lack of ability to deal with the surface irregularities of different types of stone is one example of this general failing.

Computer models can be very useful as an interface between acousticians and any clients who do not have a good grasp of theoretical acoustics. The graphics can be used to explain to a client what a proposed room could look like, or to highlight the problems which may be caused by an inappropriate choice of shape, size, or room features. There is always a danger, though, that in this world where computers are seen by many as essential tools for just about everything, the over-reliance on computer graphics and predictions will lead to a belief that it will achieve something that it will not. There are just so many parameters which contribute to the subjective character of a room, and not enough is known about their interrelationships to be able to fully programme a computer sufficiently to give accurate predictions.

In my book *Studio Monitoring Design* I quoted Ted Uzzle, from a presentation which he made to the 72nd Convention of the Audio Engineering Society in Anaheim, California, in 1982. It is worth repeating here, as it does help to put the design process in perspective: 'No sound system, no sound product, no acoustic environment can be designed by a calculator. Nor a computer, nor a cardboard slide-rule, nor a Ouija board. There are no step-by-step instructions a technician can follow; that is like Isaac Newton going to the library and asking for a book on gravity. Design work can only be done by designers, each with his own hierarchy of priorities and criteria. His three most important tools are knowledge, experience and good judgement.' Ted Uzzle, incidentally, was not anti-computerisation; in fact, his presentation was on the *progress* of computerised design. In the fifteen years separating him writing that and me writing this, there have been huge developments in the programmes for computer assisted design, but still his words hold true. We still do not know enough about subjective acoustics to effectively programme computers for design, and that is why many 'design' software packages are referred to as CAD: computer *aided* design. In many ways, it is what we do *not* know which can make acoustic design work so fascinating. There is still so much to be learned.

As I stated in the early chapters, the music recording rooms are extensions of the musical instruments, just as a guitar amplifier is an extension of the electric guitar. Studio rooms are not like control rooms. Not only are they usually *allowed* to have a sound of their own, but if that sound is deemed to be good, then they may be *desired* to have a sound of their own. I am not against the use of computers, as they are very useful research tools in the right hands, but no computer programme has yet come up with the design for a violin to equal the sound of a Stradivarius. Stradivari was a structural engineer as well as a musical instrument builder. His in-depth knowledge of the stresses and loading which his designs would demand allowed him to choose the appropriate thicknesses and cuts of the wood which were very close to their loading limits. This was doubtless of fundamental importance in his instrument making. Had Stradivari been alive today, then probably he would have shown a keen interest in computers, and would have welcomed the *extra* insight which they may have given him into his art-form; but I am sure that he would have *learned* from the computers, not *designed* with them (except perhaps as electronic sketch-pads).

Another drawback with computers is that you cannot have a conversation with them over dinner. So much of the work of a designer is involved with getting a feel from the client about the required characteristics of a proposed studio. Very frequently, prospective studio owners will only have a limited knowledge of the options available, and may not be able to fully articulate their wishes. In fact, many of their wishes may be based more on misguided beliefs or intuition rather than hard fact. It is very much a part of the job of a studio designer to discuss their feelings and requirements in depth, offering options which the studio owners may not have considered, and relating stories and anecdotes to which their clients may be able to relate more easily. In turn, the designer hears of the needs, problems, frustrations and successes of the studio owners, and these shared experiences serve to build up the designer's own store of practical experience, which will hopefully serve in the future to forewarn of problems, or to reinforce other experiences.

I suppose what irritates me most is seeing services offered such as 'send us the plans of your building, and for $X we will send to you an advanced studio design, the product of the very latest acoustic design technology', and so forth. There is no attempt to discuss the client's wants or needs. There is no personalised service, no meeting of minds, and no 'heart' in the process. In my opinion, the people providing such services do a great disservice to the whole acoustic design industry, and actually mislead many clients who fail to realise the limitations of the computer 'designs' that are being offered to them. In this technical age, however, the claims for computer designs can be deceptively seductive; but beware the pitfalls. In a somewhat back-to-front way, though, the advancement of computer modelling has very much increased the total body of acoustic understanding. Rather ironically, this has not all come from the *results* of the computer analyses, but from the great stimulation which the need for information has given to basic acoustic research. In order to make a computer program, a great number of facts and figures are necessary. Without a rigorously factual input, a computer cannot produce a factual output. In recent years, the need for more programing information has provided both the need and the funding for more basic scientific work, and this has been a great boon to the whole science of acoustics and its applied technology. In turn, due to their great analytical capability and the speed of their calculations, computers have fed back a great deal of additional information, and have brought new people into the world of acoustics who otherwise would not have taken up this elusive science. With acoustics, there is always so much more to know. In fact, the application of computers to acoustic design is rather like their application to the production of this book. Except for the photographs, all the figures in this book were produced by computers, some were even *generated* by computers, but they were all *conceived* by human brains.

10.4 Sound pulse modelling

Some of the earliest attempts to 'see' what sound was doing involved shining light beams onto mirrors fitted to resonating tuning forks, and projecting the reflexions onto a screen as the tuning fork was rotated. In the days before oscilloscopes, this was one of the only ways to 'see' sine waves. To see what was happening inside models of buildings, light was also used in a technique of sound pulse photography, as used by Sabine in 1912. The principle goes back to 1864 when Teopler showed that when parallel light rays cross a sound-field at 90° to that of the 'sound rays', the part of the sound wavefront that is met tangentially by the light rays produces two visible lines, one light, the other dark, on a projection screen behind the sound-field. In the case of 'sound pulse photography', a spark is used as an impulsive sound source, then a second spark illuminates the model room which is exposed to a photographic screen. The screen is shielded from the direct flash from the spark, in order to prevent it from washing out the faint, refracted image. The images travel along the photographic screen at the speed of sound, and the images can be sufficiently sharp to allow 1 mm wavefronts to be clearly perceived. Accurate representations of sound diffraction and diffuse reflexions can be observed, even in very small models. The progress of sound waves at different instants can be studied by altering the time intervals between the two sparks.

10.5 Light ray modelling

Light ray models have also been used in room analysis. In these cases, mirrors were used in the models, in the places where acoustically reflective surfaces were to be situated in real life. Absorbent surfaces were represented by matt black painted surfaces. This technique was mainly used when there was interest in the effects of how particular reflective surfaces in the rooms could affect any parts of the modelled room where the principle reflexions would reach. By moving the light source(s) around to all the locations of interest, the distribution of complex reflexions from an entire surface could be studied. The main limitation to this type of technique as a design tool is that the wavelengths involved bear no resemblances to the actual wavelengths of the sound in the room under study. The diffuse reflexion and diffraction expected with low frequency sound waves therefore cannot be evaluated by this process. The technique only makes possible a reasonable estimate of the behaviour of sound waves at high frequencies.

10.6 Ripple tank modelling

Another modelling technique is the use of water trays, into which are inserted profiles of the room under investigation. A simulated 'sound' can be generated by allowing a drop of water to fall into the tank in an appropriate position, and the progress of the wave can be seen by watching the ripples cross the surface of the water. The model has the advantage that the slow progress of water ripples can make the wave propagation clearly visible, but there is the drawback that the model only operates in a two-dimensional plane. 'Ripple tank' models are usually carried out in a glass bottomed tank, illuminated from below by plane-parallel light rays. This method is excellent for demonstrations, and is easily photographed. By selecting a depth of water of around 8 mm, the effects of gravity and surface tension can be effectively nullified, such that long and short wavelengths travel at more or less equal speeds, just as they do in acoustic waves. Water wave models were used by Scott Russel for acoustical investigations as far back as 1843.

10.7 Summary

In computer models the effects of all of the various physical modelling techniques can be incorporated into the programmes, but nevertheless, they are only analytical tools, just as are the physical modelling methods. For achieving the best results from computer models, it is essential that the users understand the limitations of the expressions, and of the mathematical models used by the programmes. Unfortunately, this is rarely the case. There is some very heavy mathematics involved in these models, and a *very* limited number of individuals who understand fully the mathematical *and* acoustical implications. These few people tend to reside in the academic world, and consequently may not be devoting their working lives to studio design. Neither will they necessarily have the practical design experience of professional designers. Using the whole team of specialists necessary for a full understanding of many of the computer derived implications is something which could only be afforded by governments, and not by any normal prospective studio owners.

All modelling techniques are useful methods of scientific and technical investigation and, in computer form, are very marketable items with a ready multitude of customers wanting to buy them. However, one should never expect more from the computer programme than one knows how to interpret. The graphic displays are not gospel. They are perhaps more useful in locating problems than they are in suggesting what should be done about them, but that is generally true of the whole science/art of studio design, or, for that matter, the design of any room for musical performances. The previously quoted words from Ted Uzzle still hold very true, and are likely to do so for a very long time to come.

References

1 Schroeder, M. R., 'New Method of Measuring Reverberation Time', *Journal of the Acoustic Society of America*, pp 409–411 (1965)
2 Heyser, Richard C., 'Acoustical Measurements by Time Delay Spectrometry', *Journal of the Audio Engineering Society*, Vol 15, p 370 (1967)

Glossary of terms

Below is a brief description of some of the technical terms found throughout this text. Experience has shown that although some of the terms may be familiar to many readers, they have often been misunderstood or misused. This glossary attempts to clarify the definitions of the terms in the context of their use in this book.

The decibel and sound pressure level (SPL)

Many observable physical phenomena cover a truly enormous dynamic range, and sound is no exception. The changes in pressure in the air due to the quietest of audible sounds are of the order of 20 µPa (20 micropascals), that is 0.00002 Pa, whereas those that are due to sounds on the threshold of ear-pain are of the order of 20 Pa, a ratio of one to one million. When the very loudest sounds, such as those generated by jet engines and rockets, are considered, this ratio becomes nearer to one to one thousand million! Clearly, the usual, linear number system is inefficient for an everyday description of such a wide dynamic range, so the concept of the bel was introduced to compress wide dynamic ranges into manageable numbers. The bel is simply *the logarithm of the ratio of two powers*; the decibel is one tenth of a bel.

Acoustic pressure is measured in pascals (newtons per square metre), which do not have the units of power. In order to express acoustic pressure in decibels, it is therefore necessary to square the pressure and divide it by a squared reference pressure. For convenience, the squaring of the two pressures is usually taken outside the logarithm (a useful property of logarithms); the formula for converting from acoustic pressure to decibels can then be written

$$\text{decibels} = 10 \times \log_{10} \left\{ \frac{p^2}{p_o^2} \right\} = 20 \times \log_{10} \left\{ \frac{p}{p_o} \right\}$$

where p is the acoustic pressure of interest and p_o is the reference pressure. When 20 µPa is used as the reference pressure, sound pressure expressed in decibels is referred to as *sound pressure level* (SPL). A sound pressure of 3 Pa is therefore equivalent to a sound pressure level of 103.5 decibels (dB), thus:

$$\text{SPL} = 20 \times \log_{10} \left\{ \frac{3}{20 \times 10^{-6}} \right\} = 103.5 \text{ dB}$$

The acoustic dynamic range above can be expressed in decibels as sound pressure levels of 0 dB for the quietest sounds, through 120 dB for the threshold of pain to 180 dB for the loudest sounds.

Decibels are also used to express electrical quantities such as voltages and currents, in which case the reference quantity will depend upon the application (and should *always* be stated).

When dealing with quantities that already have the units of power, such as sound power or electrical power, the squaring inside the logarithm is unnecessary and the ratio of two powers, W_1 and W_2, expressed in decibels is then

$$10 \times \log 10 \left\{ \frac{W_1}{W_2} \right\}$$

Noise weighting curves (dBA etc.)

The human ear does not have a flat frequency response; a low frequency noise will generally sound quieter than a higher frequency noise having the same sound pressure level. A measurement of sound pressure level does not, therefore, yield an accurate measurement of perceived loudness unless the frequency content of the noise is taken into account. Noise weighting curves are used to convert sound pressure level measurements into an approximation of perceived loudness, by discriminating against low and high frequency noises. The most commonly used noise weighting curve is known as A-weighting. An A-weighting curve is simply a filter with a response that rises with increasing frequency up to 2 kHz above which it falls off gently.

The frequency response of the human ear changes with changes in sound pressure level (see Figs 6.1 and 6.2), so different weighting curves are required for different levels. The dBA curve was developed for signals having a loudness below 40 phon, the dBB curve is intended for higher levels and at levels over about 80 phon, the dBC curve should be used. Other curves are also in use, such as dBD, which can be used for high level industrial noise, and dBG, which is used for infrasonic and very low frequency noise assessments. The responses of the various weighting curves are shown in Fig. A1.

The widespread use of the dBA curve for the assessment of noise can give rise to poor results in situations when another weighting curve is more appropriate. For example, Fig. A2 shows the dBA curve superimposed on the inverse of the equal loudness contours – similar to those in Fig. 6.2, but upside down. Only at about 1 kHz and 6 kHz do the curves agree. Between 3 kHz and 4 kHz, errors of up to 10 dB can be seen, and at low frequencies, the A weighting curve can over-assess or under-assess noise nuisance levels by up to 20 dB, depending upon level. The dBA curve is often used at relatively high levels; a purpose for which it was never intended and is not suited.

In any case, noise weighting should only be applied when one requires an approximation to the perceived loudness of a sound; it is therefore of most

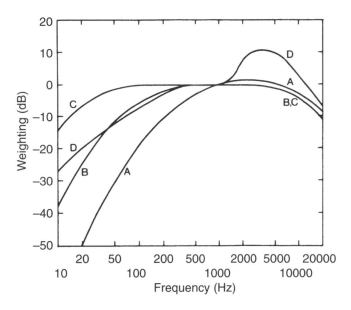

Figure A1 A-, B-, C- and D-weighting curves for sound level meters

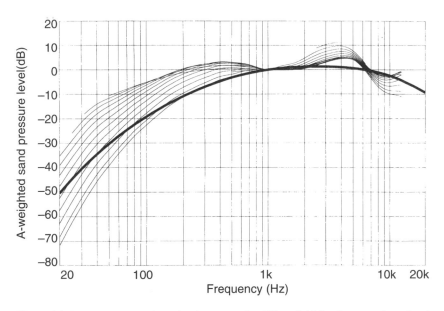

Figure A2 Inverse equal loudness level contours for diffuse field for the range from threshold to 120 phon (thin lines) and A-weighting (thick line). The above figure was taken from 'On the Use of A-weighted Level for Prediction of Loudness Level', by Henrik Møller and Morten Lydolf, of Aalborg University, Denmark. The paper was presented to the 8th International Meeting on Low Frequency Noise and Vibration, in Gothenburg, Sweden, in June 1997. The work is published in the Proceedings, by Multi-Science Publishing, UK

use in noise assessment. Noise weighting should never be applied when absolute values of sound pressure are required; in the measurement of loud-speaker frequency response for example.

Frequency

Frequency is a measure of how often a given event is repeated. In acoustics, the frequency of a sound is the number of cycles of alternate positive and negative pressures that occur in one second. The units of frequency are the hertz (Hz) or, more traditionally, cycles per second (cps).

Audio frequency ranges

The range of frequencies over which the human ear is sensitive is usually considered to be from 20 Hz to 20 kHz. For convenience, it is desirable to split this range into a number of smaller ranges. A number of commonly used frequency ranges are listed below. The span of frequencies quoted for each range should not be treated as standard, they are included as a 'rough' guide only.

Name	Frequency range
Infrasonic	0–20 Hz
Very low	15 Hz–50 Hz
Low	20 Hz–250 Hz
Lower mid	200 Hz–500 Hz
Mid	250 Hz–5k Hz
Upper mid	2k Hz–6k Hz
High	5k Hz–20 kHz
Very high	15 kHz–25 kHz
Ultrasonic	20 kHz–∞

The sine wave and its frequency content

The sine wave is a graph of the value of a single frequency signal against time. Strictly speaking, for a signal to consist of a single frequency, the sine wave must have existed for all time, as any change to the amplitude of the signal, such as during a switch-on or switch-off, gives rise to the generation of other frequencies; this has important implications for audio. Most audio signals contain pseudo-steady-state sounds, such as notes played on an instrument. When these sounds are reproduced by an imperfect audio system, the excitation of any resonances in the reproduction chain will depend upon the frequency content of the signal. During a long note, the signal may be dominated by a few discrete frequencies, such as a set of harmonics, and the chances of resonances being excited are slim. However, during the start and stop of the note, a range of frequencies are produced, above and below those of the steady-state signal, and the chances of resonances being excited are increased. This phenomenon leads to the apparent pitch of the note being 'pulled' towards the frequency of any nearby resonance during the start, and particularly the end, of the note.

Standing waves and resonances

Standing waves occur whenever two or more waves having the same frequency and type pass through the same point. The resultant spatial interference pattern, which consists of regions of high and low amplitude, is 'fixed' in space even though the waves themselves may be travelling.

Resonant standing waves only occur when a standing wave pattern is set up by the interference between a wave and its reflections from two or more surfaces *and* when the wave travels from a point, via the surfaces, back to that point, it is travelling in the original direction *and* when the distance travelled by this wave is equal to an exact number of wavelengths. The returning wave then reinforces itself, and if losses are low, the standing wave field becomes resonant.

The simplest resonant standing wave to visualise is that set up between two parallel walls spaced one half of a wavelength apart. A wave travelling from a point towards one of the walls is reflected back towards the other wall from which it is reflected back again in the original direction. As the distance between the walls is one half of a wavelength, the total distance travelled by the wave on return to the point is one wavelength; the wave then travels away from the point with exactly the right phase to reinforce the next cycle of the wave. If the frequency of the wave or the distance between the walls is changed, a standing wave pattern will still exist between the walls but resonance will not occur.

It should be stressed that standing waves *always* exist when like waves interfere, whether a resonance situation occurs or not, and that the common usage of the term 'standing wave' to describe only resonant conditions is both erroneous and misleading.

Damping

Damping refers to any mechanism which causes an oscillating system to lose energy. Damping of acoustic waves can result from the frictional losses associated with the propagation of sound through porous materials, the radiation of sound power, or causing a structure with internal losses to vibrate.

Microphone directivity patterns

Most microphones consist of a small diaphragm which moves in reponse to changes in the pressure exerted on it by a sound wave; the diaphragm motion is then detected and converted into an electrical signal

The simplest form of microphone is one that has only one side of the diaphragm exposed to the sound field. If the diaphragm is sufficiently small, such a microphone will respond equally to sounds from all directions and is termed omnidirectional

A microphone which has both sides of the diaphragm open to the sound field will only detect the difference between the pressures on the two sides. When a sound wave is incident from a direction normal to the diaphragm, there will be a short delay between the pressure on the incident side and that on the far side, and the microphone responds to the resultant pressure difference. When a sound wave is incident from a direction in line with the diaphragm, the same pressure is exerted on both sides of the diaphragm and the sound is not detected. This arrangement results in a 'figure-of-eight' or dipole directivity pattern.

If an omnidirectional microphone element and a figure-of-eight microphone element are mounted close together and their outputs summed, the resultant directivity pattern will lie between the two extremes of omnidirectional and figure-of-eight patterns. If the sensitivities of the two elements are the same, the combined directivity pattern is known as cardioid, because of its heart shape. Various other patterns, such as hypercardioid and supercardioid are achieved by varying the relative sensitivities of the omnidirectional and figure-of-eight elements. Similar directivity patterns can be realised using only one microphone element. The microphone diaphragm is mounted at one end of a short tube, and the delay introduced to one side of the diaphragm by the tube gives rise to an approximation to a cardioid directivity pattern. More complex directivity patterns can be achieved by using more than two elements; the SoundField microphone, for example, has six elements which can be combined in a variety of ways.

Psychoacoustics

Unlike the related discipline of acoustics, which is concerned with the physics of sound, psychoacoustics is the science of the perception of sound, particularly by humans. The stereo illusion, the cocktail party effect and the perception of pitch are all examples of psychacoustic phenomena.

Objective and subjective assessment

In acoustics in general, and in audio in particular, there is some disagreement between that which our measurements tell us and that which we hear. In audio, objective assessment involves measuring the performance of a piece of equipment using instruments and comparing this performance with a desired specification. Subjective assessment, on the other hand, involves auditioning the equipment under carefully controlled conditions and assessing particular aspects of the sounds that are heard. The successful assessment of the quality or suitability of a piece of audio equipment therefore ideally needs both approaches. Objective assessment is more easily carried out in the laboratory, or during production runs, than subjective assessment; to make a reliable and repeatable subjective assessment usually requires the ears of a number of subjects and hence, often, a large amount of time.

List of abbreviations

SPL   sound pressure level
dB   decibel
Lf   low-frequency
Mf   mid-frequency
Hf   high-frequency
ETC   energy/time curve
Hz   hertz

Index